10 Minnesota MCA Grade 6 Math Practice Tests

The Ultimate Test Prep Collection with Answer Explanations

Dr. A. Nazari

10 Practice Tests

The Grand Championship Collection

Welcome, future Math Champion!

*You hold the **ultimate collection** —
ten full-length practice tests designed to take you
from first attempt to **complete mastery**.*

Conquer every Grade 6 topic

Build unshakeable confidence

Rise from Bronze to Gold to Champion

Arrive at test day fully prepared

The championship begins now.

❝** Ten tests may seem like a marathon, but champions are made one step at a time. Trust the process! **❞

♛ The Champion's Path ♛

Your 4-phase journey from Bronze to Champion

🥉 Bronze Round (Tests 1–3)

Your warm-up matches. Take these **untimed** to learn the format and set your baseline. Read the answer explanations after each test — this is where you build your foundation.

🥈 Silver Round (Tests 4–6)

Set a timer for **75 minutes**. Focus on the topics that tripped you up in Bronze. Practice showing your work on every problem. Your accuracy should be climbing.

🥇 Gold Round (Tests 7–9)

Full timed conditions (**60 minutes**). Simulate the real exam environment. Review only the questions you missed — targeted practice is the key to gold.

👑 Championship Final (Test 10)

Your final match. Full exam conditions — timed, quiet, no breaks. This is your victory lap. Show yourself how far you've come!

🏆 Your Championship Kit

- 🥇 **10 Full-Length Practice Tests** — every Grade 6 topic
- 🥇 **Formula Reference Sheet**
- 🥇 **Complete Answer Key** with explanations
- 🥇 **Championship Scoreboard** to track your rise

Champion's Tip: Space your tests 2–3 days apart. Use the days in between for targeted review By Test 10, you'll be amazed at your transformation.

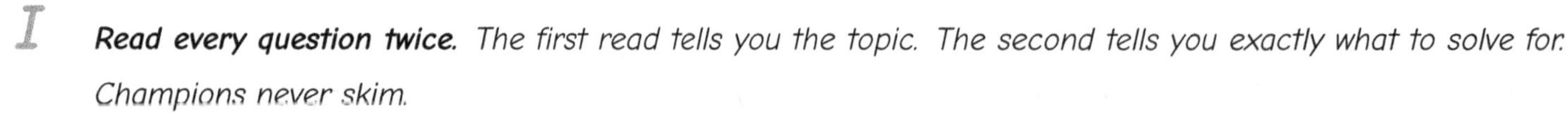

The Champion's Playbook

Seven rules that separate champions from the rest

I **Read every question twice.** The first read tells you the topic. The second tells you exactly what to solve for. Champions never skim.

II **Mark the clues.** Circle key numbers, underline the question, and cross out information that's just there to distract you.

III **Choose your strategy.** Before touching pencil to paper, decide: Am I setting up a ratio? Solving an equation? Finding area? Name the approach.

IV **Solve, then match.** For multiple choice — work the problem on scratch paper first, then find your answer among the choices.

V **Eliminate and conquer.** Cross out obviously wrong answers. If you're left with two, you've already doubled your odds. Make an educated pick.

VI **Estimate to verify.** After solving, ask: "Is this answer reasonable?" A quick mental estimate catches most calculation errors.

VII **Leave nothing blank.** Even a well-reasoned guess is worth more than an empty space. Use partial work to support your answer.

⏱ Timing Mastery

Tests 1–3: **Untimed** (build foundation) › Tests 4–6: **75 min** (build speed) › Tests 7–10: **60 min** (championship conditions)

★ Grade 6 Championship Topics

🎖 Ratios & Proportions   🎖 Integers & Rational Numbers   🎖 Expressions & Equations

🎖 Geometry & Measurement   🎖 Statistics & Data Analysis

*A true champion isn't someone who never makes mistakes — it's someone who learns from **every single one**. After each test, review your errors carefully. That's where the real growth happens.*

Get Online

Find more at
ViewMath.com/MN-Grade6

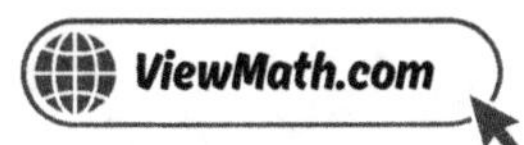

👑 The Champion's Toolkit 👑

Prepare your workspace before each championship round

🏆 Required Equipment

Sharpened Pencils — Two #2 pencils — champions always have a backup

Quality Eraser — A clean, soft eraser that won't smudge your work

Scratch Paper — Blank paper for calculations, diagrams, and number lines

Ruler — Essential for geometry and coordinate plane questions

Timer — Begin using from the Silver Round onward

Quiet Workspace — A calm, well-lit area free from distractions

🏅 Permitted in Competition

- ✔ Pencil and eraser
- ✔ Scratch paper (provided)
- ✔ Ruler (if specified)
- ✔ Formula reference in this book

🚫 Not Permitted

- ✘ Calculators
- ✘ Electronic devices
- ✘ Textbooks or notes
- ✘ Outside help

Find more at
ViewMath.com/MN-Grade6

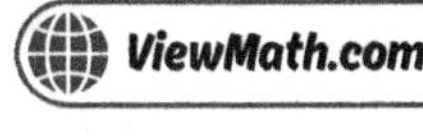

Formula Reference Sheet

Area Formulas

Rectangle	$A = l \times w$
Parallelogram	$A = b \times h$
Triangle	$A = \dfrac{1}{2} \times b \times h$
Trapezoid	$A = \dfrac{1}{2}(b_1 + b_2) \times h$

Volume

Rectangular $V = l \times w \times h$

Prism

Surface Area

Find the area of each face, then add them all up.

Rectangular Prism:

$SA = 2lw + 2lh + 2wh$

Order of Operations

P Parentheses first

E Exponents

M/D Multiply & Divide (left to right)

A/S Add & Subtract (left to right)

Ratios & Percents

Ratio: $a : b$ or $\dfrac{a}{b}$

Unit rate: amount per 1 unit

Percent: a ratio out of 100

Part = Percent $\times$ Whole

Integers & Absolute Value

Integers:

$\ldots, -3, -2, -1, 0, 1, 2, 3, \ldots$

$|-5| = 5 \quad |5| = 5$

Absolute value = distance from 0

X¹ Expressions & Equations

Exponent: $3^4 = 3 \times 3 \times 3 \times 3 = 81$

Variable: a letter that stands for a number

Equation: two expressions joined by $=$

Inequality: uses $<, >, \leq, \geq$

Coordinate Plane

Ordered pair: (x, y)

x-axis: horizontal y-axis: vertical

Origin: $(0, 0)$

Four quadrants (I, II, III, IV)

Statistics

Mean: sum of values $\div$ count

Median: middle value (sorted)

Range: max $-$ min

Championship Scoreboard

Track your rise through every championship round

Champion's Name: ___________________________

Round	Tier	Date	Score	Rating
1	Bronze		/	
2	Bronze		/	
3	Bronze		/	
4	Silver		/	
5	Silver		/	
6	Silver		/	
7	Gold		/	
8	Gold		/	
9	Gold		/	
10	♛		/	

♛ Champion's Reflection

My strongest topics (where I consistently score well):

____

____

Topics I improved on the most from Bronze to Gold:

____

____

My score trend (Bronze avg → Gold avg → Championship):

____

____

One strategy that helped me improve the most:

____

____

My confidence level for the real test (1–10): _________ / 10

Find more at
ViewMath.com/MN-Grade6

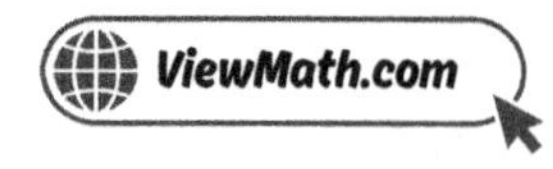

★ Table of Contents ★

Here's what we'll explore together!

 Let's learn and have fun!

Practice Test 1

📋 *30 Questions*

✏️ Before You Start ✏️

- ✓ **Read each question carefully** before choosing your answer.
- ✓ **Show your work** on scratch paper when you need to.
- ✓ **Skip hard questions** and come back to them later.
- ✓ **Check your answers** when you're done.
- ✓ **Take your time** — there's no rush!

⭐ You've Got This! ⭐

Do your best and show what you know!

1. A garden has flowers and vegetables in a ratio of $4 : 1$. There are 20 flowers. How many vegetables are there?

Your Answer:

2. The graph shows the relationship between hours worked and money earned.

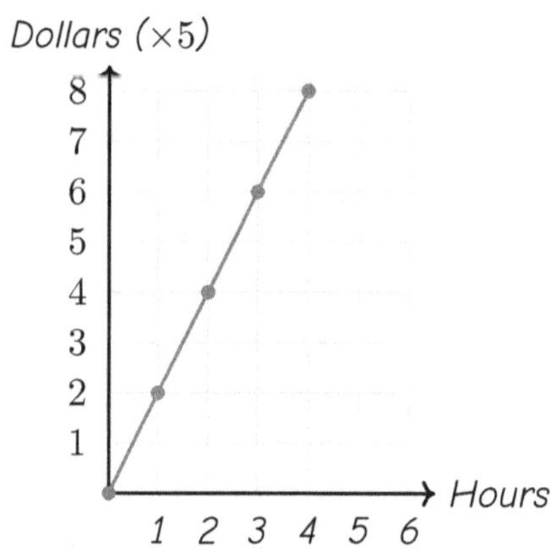

How much money is earned in 6 hours?

A) $30

B) $50

C) $60

D) $40

3. The table and partial graph below show a ratio relationship.

x	y
1	4
2	8
3	?

What is the missing value of y when $x = 3$?

(A) 10

(B) 12

(C) 14

(D) 16

4. Write 0.15 as a percent.

(A) 1.5%

(B) 15%

(C) 150%

(D) 0.15%

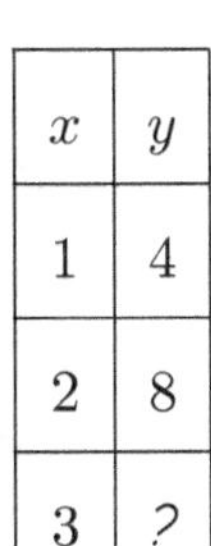

5. The table compares four payment methods.

Method	Uses Your Own Money?	Can Charge Interest?
Cash	Yes	No
Check	Yes	No
Debit Card	Yes	No
Credit Card	No	Yes

Based on the table, which statement is true?

(A) A check borrows money from the bank

(B) A debit card can charge you interest

(C) Only a credit card involves borrowing money

(D) Cash and credit cards work the same way

6. Two friends each earn money mowing lawns.

Ella: $y = 15x$ **Jake:** $y = 12x + 10$

Who has a proportional relationship between lawns mowed (x) and earnings (y)? After mowing 6 lawns, who earns more?

Your Answer

7. The number line below shows equal jumps from 0 to $\dfrac{5}{6}$.

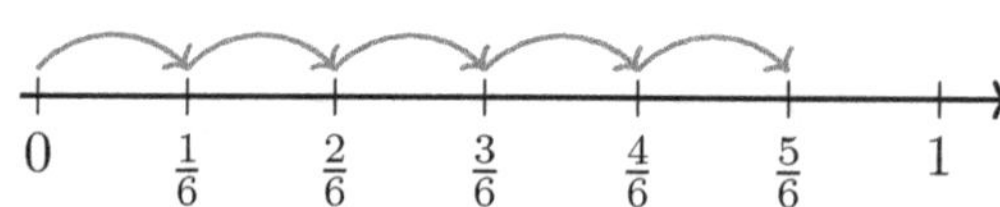

Which division problem does the number line represent?

(A) $\dfrac{5}{6} \div \dfrac{1}{6} = 5$

(B) $\dfrac{1}{6} \div \dfrac{5}{6} = 5$

(C) $\dfrac{5}{6} \times \dfrac{1}{6} = \dfrac{5}{36}$

(D) $\dfrac{5}{6} \div 5 = \dfrac{1}{6}$

8. A water jug holds 5.4 liters. How many liters are in 6 jugs?

   (A) 32.4

   (B) 324

   (C) 3.24

   (D) 30.4

9. Find $|-9| + |-6|$.

Your Answer:

10. The point $(-7, 3)$ is reflected across the x-axis. Write the coordinates of the reflected point.

Your Answer:

11. A store charges the prices shown below. Which expression gives the total cost for s sandwiches and d drinks?

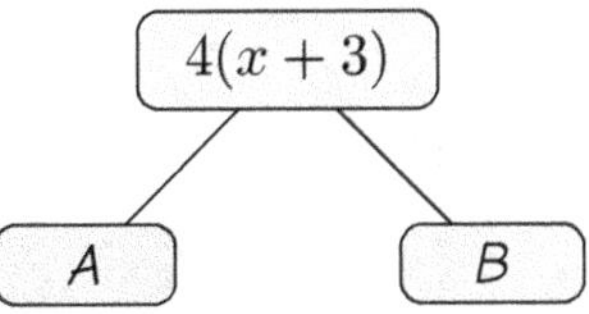

   (A) $6 + 2$

   (B) $6s + 2d$

   (C) $8sd$

   (D) $2s + 6d$

12. A student made a factor tree for the expression $4(x+3)$, shown below. Fill in the missing pieces labeled A and B.

Your Answer:

Find more at
ViewMath.com/MN-Grade6

ViewMath.com

13. Evaluate $3(p + 2) - p$ when $p = 8$.

Your Answer:

14. Which expression is equivalent to $3(x + 5)$?

A) $3x + 5$

B) $3x + 15$

C) $8x$

D) $3x + 8$

15. Write a situation that could be modeled by the expression $5n + 3$.

Your Answer:

16. The temperature stayed below 0 degrees. Which inequality represents the temperature t?

A) $t > 0$

B) $t < 0$

C) $t \leq 0$

D) $t \geq 0$

17. A student says the graph of $x < 8$ and the graph of $x \leq 8$ look exactly the same. Is this correct?

Your Answer:

18. Complete the table for $y = 4x - 1$:

$x = 1$: $y = ?$ $x = 3$: $y = ?$ $x = 5$: $y = ?$

Your Answer:

Find more at
ViewMath.com/MN-Grade6

ViewMath.com

19. *A triangular sail has a base of 3 m and a height of 7 m. How much fabric is needed to make 2 identical sails?*

(A) $10.5\ m^2$

(B) $42\ m^2$

(C) $21\ m^2$

(D) $20\ m^2$

20. *Two boxes have the same volume. Box A is $6 \times 4 \times 5$. Box B is $10 \times 3 \times h$. What is the height h of Box B?*

(A) 2

(B) 4

(C) 6

(D) 8

21. *A right triangle has vertices $(-4, 0)$, $(2, 0)$, and $(-4, 5)$. What is the area?*

(A) 30 *square units*

(B) 15 *square units*

(C) 11 *square units*

(D) 20 *square units*

22. *A gift box is 12 in long, 8 in wide, and 5 in tall. How much wrapping paper is needed to cover all sides?*

(A) $392\ in^2$

(B) $480\ in^2$

(C) $296\ in^2$

(D) $196\ in^2$

23. *What is the area of a circle with radius 5 inches? Use $\pi \approx 3.14$.*

(A) $15.7\ in^2$

(B) $31.4\ in^2$

(C) $78.5\ in^2$

(D) $157\ in^2$

24. If you ask 20 classmates, "How many pets do you have?" and you get the answers $0, 1, 1, 2, 0, 3, 1, 2, 0, 1, 4, 2, 1, 0, 2, 1, 3, 1, 0, 2,$ what does this tell you about the question?

(A) It is not a statistical question because some answers are 0.

(B) It is statistical because the answers vary.

(C) It is not statistical because the answers are whole numbers.

(D) It is statistical because there are 20 answers.

25. Data: $5, 10, 15, 20, 25, 30, 35.$ What is the range?

(A) 5

(B) 15

(C) 20

(D) 30

26. A dot plot shows hours of TV watched daily: 0 (3 dots), 1 (6 dots), 2 (5 dots), 3 (4 dots), 4 (1 dot), 8 (1 dot).

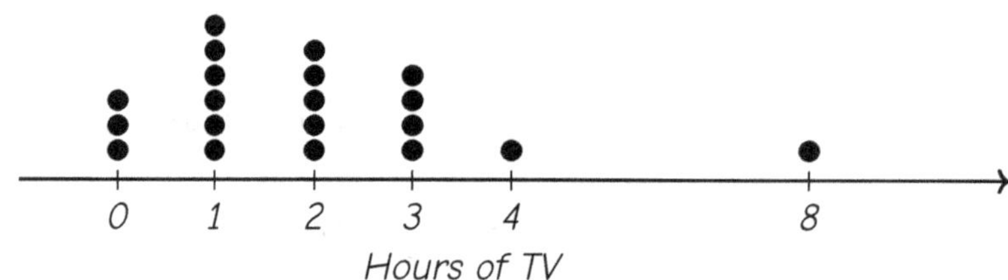

What is the mode, and which value appears to be an outlier?

(A) Mode = 1 hour; outlier = 0

(B) Mode = 1 hour; outlier = 8 hours

(C) Mode = 2 hours; outlier = 8 hours

(D) Mode = 3 hours; outlier = 4 hours

27. Which part of the box plot represents the lowest 25% of the data?

(A) The box

(B) The left whisker (from min to Q1)

(C) The right whisker (from Q3 to max)

(D) The line at the median

Find more at
ViewMath.com/MN-Grade6

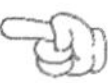

ViewMath.com

28. A student says "The data set with the larger range is always more spread out than the one with the smaller range." Is this always true? Explain.

Your Answer:

29. A standard number cube is rolled. What is the probability of rolling a number less than or equal to 3?

(A) $\frac{1}{3}$

(B) $\frac{3}{6}$

(C) $\frac{2}{6}$

(D) $\frac{4}{6}$

30. A family earns $3,000 per month. Their circle graph shows 15% goes to entertainment. How many dollars go to entertainment each month?

Your Answer:

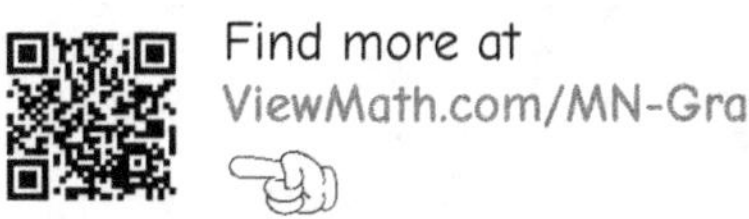

 # End of Practice Test 1 

Great job finishing the test!

 ## My Score

I got _____________ out of 30 questions right.

*Check your answers in the **Answer Key** at the back of the book.*

💡 *Review any questions you missed. That's how we learn!*

📊 Check Your Score Online!

*Visit **ViewMath Academy** to enter your answers and see which topics you need to review. You can also explore lessons, take quizzes, track your scores, and save your progress!*

viewmath.com/score/6.1.MN.16

Or go to viewmath.com/score and enter code: 6.1.MN.16

Practice Test 2

 30 Questions

✏️ Before You Start ✏️

- ✔ **Read each question carefully** before choosing your answer.
- ✔ **Show your work** on scratch paper when you need to.
- ✔ **Skip hard questions** and come back to them later.
- ✔ **Check your answers** when you're done.
- ✔ **Take your time** — there's no rush!

⭐ You've Got This! ⭐

Do your best and show what you know!

1. Look at the tape diagram below.

Boys: □□□□
Girls: □□□□□□

There are 24 girls. How many boys are there?

(A) 4

(B) 12

(C) 16

(D) 20

2. Complete the ratio table below. The ratio of markers to crayons is $4 : 6$.

Markers	Crayons
4	6
8	?
?	24
20	?

Find all three missing values.

Your Answer

Find more at
ViewMath.com/MN-Grade6

3. Two runners' distances are graphed below.

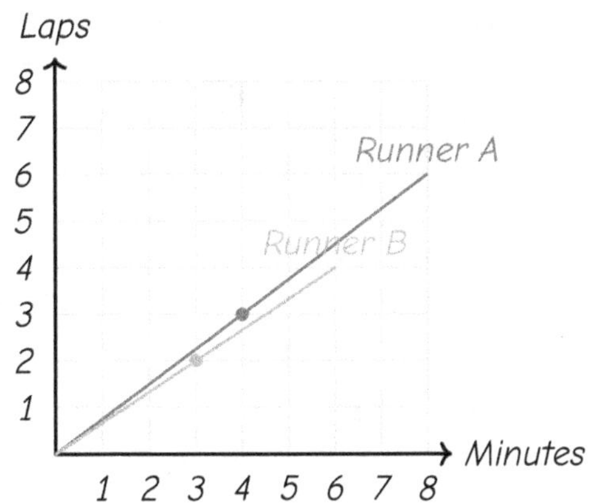

Part A: What is each runner's ratio of laps to minutes?

Part B: Who runs faster? Explain using the graph.

Your Answer:

4. A meal costs $25. You leave a 20% tip. How much is the tip?

(A) $3

(B) $4

(C) $5

(D) $6

5. Olivia earns $240 per month from her part-time job. She saves 20% and spends 10% on school supplies. How much money is left for other expenses?

(A) $72

(B) $144

(C) $168

(D) $192

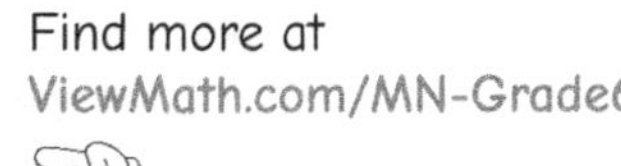

6. A recipe uses 2 cups of flour for every 3 cups of sugar. If you use 6 cups of flour, how many cups of sugar do you need?

    (A) 7

    (C) 9

    (B) 8

    (D) 12

7. What is $\dfrac{3}{5} \div \dfrac{1}{5}$?

    (A) $\dfrac{3}{25}$

    (C) $\dfrac{1}{3}$

    (B) 3

    (D) 15

8. A book costs \$12.95 and a bookmark costs \$3.48. Jenna buys one of each. How much does she spend in all?

    Your Answer:

9. What is $|-12|$?

    (A) -12

    (C) 0

    (B) 12

    (D) $-(-(-12))$

10. Explain why the point $(9, 0)$ is not in any quadrant.

    Your Answer:

11. A baker makes c cupcakes. She packs them equally into 6 boxes. Which expression shows the number in each box?

(A) $c - 6$

(B) $6c$

(C) $c \div 6$

(D) $c + 6$

12. In the expression $5(y + 8)$, name the two factors.

Your Answer:

13. The triangle below has a base of $b = 10$ cm and a height of $h = 6$ cm. Use the formula $A = \frac{1}{2} \times b \times h$ to find its area.

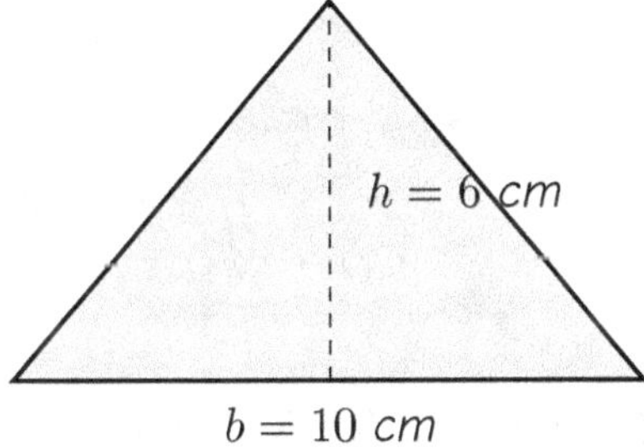

Your Answer:

14. Simplify: $3a + 7 + 5a - 2a + 3$

Your Answer:

15. A movie ticket costs \$9. Your group also shares a \$6 bucket of popcorn. Which expression represents the total cost for p people?

(A) $9p + 6$

(B) $9 + 6p$

(C) $15p$

(D) $9p - 6$

16. Which inequality represents "a number y is fewer than 20"?

(A) $y > 20$

(B) $y \geq 20$

(C) $y < 20$

(D) $y \leq 20$

17. What is the difference between the graphs of $x > 5$ and $x \geq 5$?

(A) They shade in different directions

(B) $x > 5$ uses an open circle; $x \geq 5$ uses a closed circle

(C) $x > 5$ shades right; $x \geq 5$ shades left

(D) There is no difference

18. Which equation represents this relationship: "The number of legs L is 4 times the number of dogs d"?

(A) $d = 4L$

(B) $L = d + 4$

(C) $L = 4d$

(D) $L = d \div 4$

19. A right triangle has legs of 9 in and 12 in. What is its area?

Your Answer

20. A rectangular prism has a volume of 120 cm^3. Its length is 10 cm and width is 4 cm. What is the height?

(A) 3 cm

(B) 12 cm

(C) 6 cm

(D) 30 cm

Find more at
ViewMath.com/MN-Grade6

21. A rectangle on the coordinate plane has area 54 square units. Two of its vertices are at $(-3, 2)$ and $(6, 2)$. What is the width of the rectangle?

(A) 6 units

(B) 9 units

(C) 3 units

(D) 18 units

22. A classroom has dimensions 10 m by 8 m by 3 m. What is the total surface area of the walls, floor, and ceiling?

Your Answer:

23. A circle has a diameter of 12 m. What is its area? Use $\pi \approx 3.14$.

(A) 37.68 m^2

(B) 113.04 m^2

(C) 226.08 m^2

(D) 452.16 m^2

24. A student says, "Any question with a number for an answer is statistical." Is the student correct? Explain.

Your Answer:

25. A number line below shows the positions of Q1, the median, and Q3 for a data set.

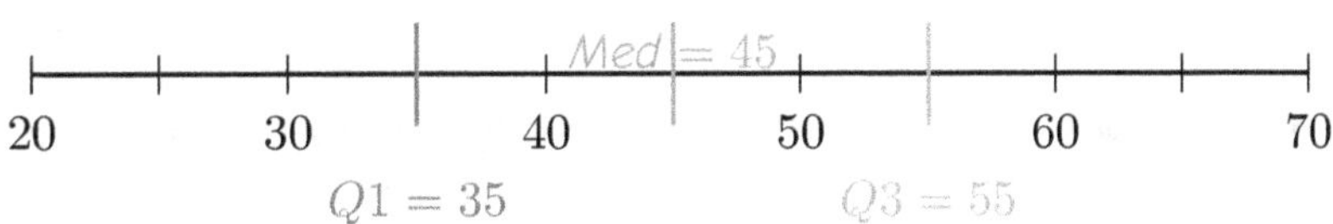

What is the IQR?

(A) 10

(B) 15

(C) 20

(D) 35

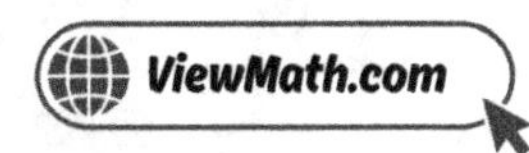

Find more at
ViewMath.com/MN-Grade6

Get Online

ViewMath.com

26. The frequency table below shows the results of a school survey about how students get to school.

Transportation	Frequency
Bus	45
Car	30
Walk	15
Bike	10

Find the total number of students surveyed, the mode, and the percent who walk to school.

Your Answer:

27. Which five values make up the five-number summary?

(A) Mean, median, mode, range, IQR

(B) Minimum, Q1, median, Q3, maximum

(C) Q1, Q2, Q3, Q4, Q5

(D) Mean, Q1, Q3, range, MAD

28. A coach records sprint times (seconds) for two athletes over 10 races. Athlete A: median $= 12.5$, range $= 1.2$. Athlete B: median $= 12.5$, range $= 3.0$. Which athlete is more consistent?

(A) Athlete A

(B) Athlete B

(C) They are equally consistent.

(D) Cannot be determined without the IQR.

29. The probability scale below shows four events. Which event has a probability closest to 0.75?

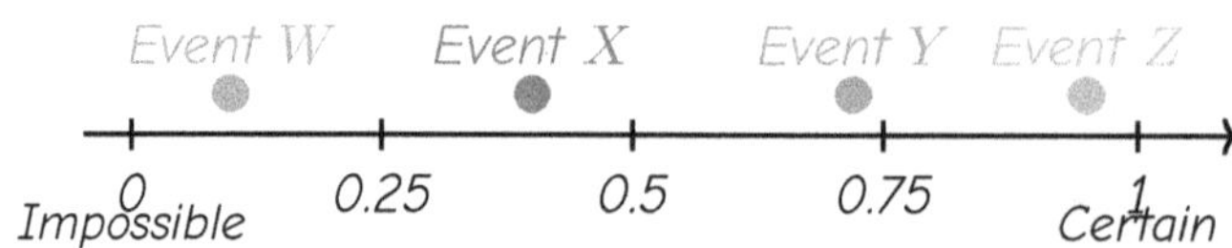

(A) Event W

(B) Event X

(C) Event Y

(D) Event Z

Find more at
ViewMath.com/MN-Grade6

ViewMath.com

30. The circle graph below shows how a family spends its monthly income.

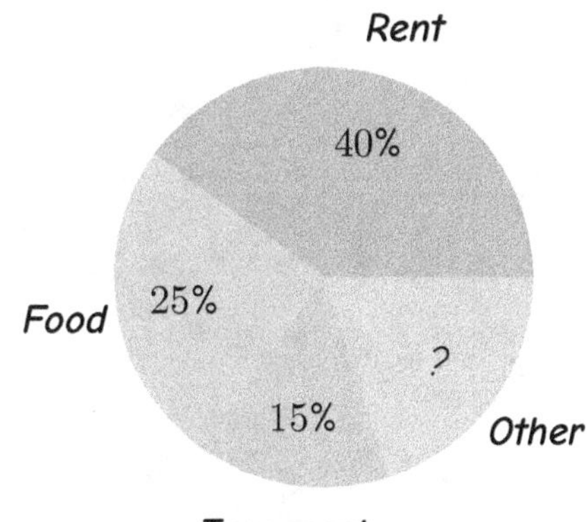

What percent of the family's income goes to the **Other** category?

(A) 10%

(B) 15%

(C) 20%

(D) 25%

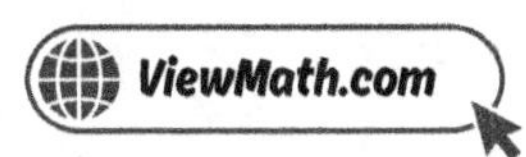

⭐ End of Practice Test 2 ⭐

Great job finishing the test!

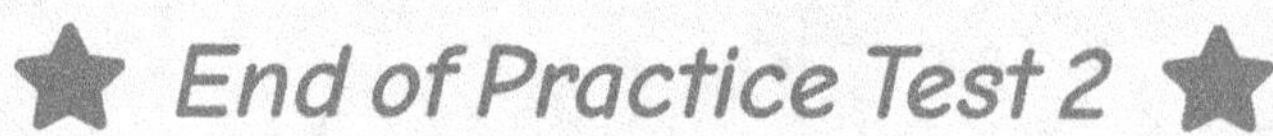

✅ My Score

I got _____________ out of 30 questions right.

*Check your answers in the **Answer Key** at the back of the book.*

💡 *Review any questions you missed. That's how we learn!*

📊 Check Your Score Online!

Visit **ViewMath Academy** to enter your answers and see which topics you need to review. You can also explore lessons, take quizzes, track your scores, and save your progress!

viewmath.com/score/6.1.MN.17

Or go to viewmath.com/score and enter code: 6.1.MN.17

Practice Test 3

🗒 30 Questions

✏️ *Before You Start* ✏️

- ✓ **Read each question carefully** before choosing your answer.
- ✓ **Show your work** on scratch paper when you need to.
- ✓ **Skip hard questions** and come back to them later.
- ✓ **Check your answers** when you're done.
- ✓ **Take your time** — there's no rush!

⭐ You've Got This! ⭐

Do your best and show what you know!

1. A baker uses 2 eggs **per** 3 cups of flour. Which ratio represents flour to eggs?

   (A) $2:3$ (B) $3:5$

   (C) $3:2$ (D) $2:5$

2. A recipe uses 4 eggs for every 6 cups of flour. How many eggs are needed for 18 cups of flour?

   (A) 6   (B) 10

   (C) 12   (D) 8

3. A ratio graph passes through $(6, 2)$. What is the value of y when $x = 15$?

   (A) 3   (B) 5

   (C) 7   (D) 10

4. A pair of shoes costs \$80. They are 15% off. What is the sale price?

   (A) \$60   (B) \$65

   (C) \$68   (D) \$72

5. Emma owes \$500 on a credit card with 18% annual interest. She makes no payments for one year. What is her new balance?

   (A) \$518   (B) \$540

   (C) \$590   (D) \$680

Find more at
ViewMath.com/MN-Grade6

6. Which equation represents a **proportional** relationship?

(A) $y = 4x + 1$
(B) $y = 7x$
(C) $y = x - 3$
(D) $y = 2x + 5$

7. What is $\frac{1}{3} \div \frac{2}{3}$?

(A) $\frac{1}{2}$
(B) $\frac{2}{9}$
(C) $\frac{2}{3}$
(D) 2

8. What is 1.2×0.3?

(A) 3.6
(B) 0.36
(C) 36
(D) 0.036

9. Which number has an absolute value of 6?

(A) 6 only
(B) -6 only
(C) Both 6 and -6
(D) No number has an absolute value of 6

10. Which sign convention describes all points in Quadrant II?

(A) $(+, +)$
(B) $(-, +)$
(C) $(-, -)$
(D) $(+, -)$

11. Sam has x stickers. He gives away 6. Which expression shows how many stickers Sam has now?

(A) $x + 6$
(B) $6x$
(C) $6 - x$
(D) $x - 6$

Find more at
ViewMath.com/MN-Grade6

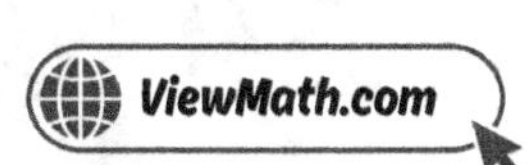

12. How many terms does the expression $4(a + b)$ have when written in its factored form?

Your Answer

13. Evaluate $k^2 + k$ when $k = 3$.

  (A) 6   (B) 9

  (C) 12   (D) 15

14. Use the distributive property to expand $8(y + 3)$.

Your Answer

15. In the formula $d = 60t$, what could d and t represent?

  (A) d = days, t = hours   (B) d = distance in miles, t = time in hours

  (C) d = dollars, t = tax rate   (D) d = degrees, t = temperature

16. A library book can be checked out for at most 21 days. Which inequality models the number of days d?

  (A) $d < 21$ (B) $d > 21$

  (C) $d \leq 21$ (D) $d \geq 21$

17. *Which inequality does the number line below represent?*

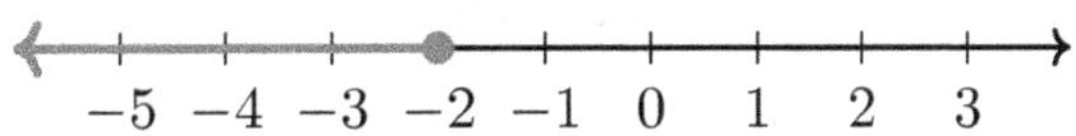

(A) $x > -2$

(B) $x < -2$

(C) $x \geq -2$

(D) $x \leq -2$

18. *In $c = 7p$, where c is total cost and p is the number of pizzas, what is the cost of 6 pizzas?*

(A) $13

(B) $36

(C) $42

(D) $67

19. *A triangle has base 15 cm and height 4 cm. Sam says the area is 60 cm^2. What mistake did Sam make?*

(A) *He used the wrong formula entirely.*

(B) *He forgot to divide by 2.*

(C) *He added instead of multiplied.*

(D) *He subtracted the height from the base.*

20. *A toy box is 3 ft long, 2 ft wide, and 2 ft tall. What is the volume?*

(A) 7 ft^3

(B) 12 ft^3

(C) 6 ft^3

(D) 24 ft^3

21. *A rectangle has vertices $(0,0)$, $(7,0)$, $(7,3)$, and $(0,3)$. A right triangle is cut from one corner with legs of 2 and 3. What is the area of the remaining shape?*

(A) 18 square units

(B) 21 square units

(C) 24 square units

(D) 15 square units

Find more at
ViewMath.com/MN-Grade6

ViewMath.com

22. *The net below shows a triangular prism. The triangles are right triangles with legs 3 cm and 4 cm. The three rectangles have widths matching the triangle sides and a shared length of 8 cm. Find the total surface area.*

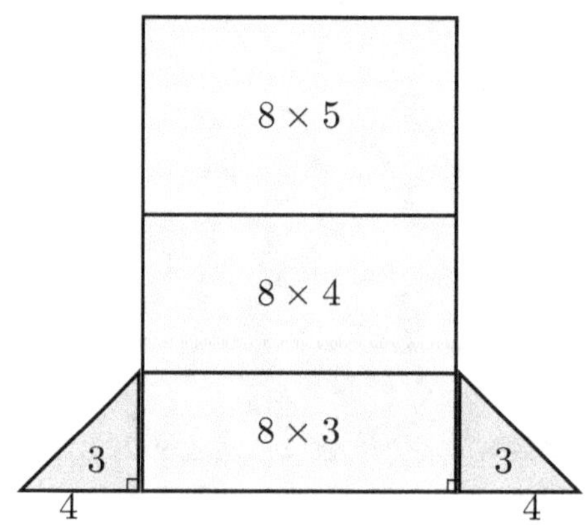

Your Answer

23. *What is the perimeter of a semicircle with a radius of 7 cm? (Include the straight edge.) Use $\pi \approx 3.14$.*

(A) 21.98 cm

(B) 35.98 cm

(C) 43.96 cm

(D) 57.96 cm

24. *Which question would produce data where every answer is the same?*

(A) How many hours do you watch TV each day?

(B) How many feet are in one yard?

(C) What is your favorite sport?

(D) How far can you throw a ball?

25. *Which statement about MAD is true?*

(A) MAD measures the middle value of the data.

(B) MAD tells you the average distance of each value from the mean.

(C) MAD is always larger than the range.

(D) MAD equals the range divided by 2.

Find more at
ViewMath.com/MN-Grade6

ViewMath.com

26. *A dot plot of daily steps:* 3000 (1), 4000 (2), 5000 (6), 6000 (4), 7000 (2). *What is the mode? What is the range?*

Your Answer:

27. *Data:* 1, 3, 5, 7, 9, 11, 13, 15, 17, 19. *What is the IQR?*

(A) 8

(B) 10

(C) 14

(D) 18

28. *A data set has a range of 40 but an IQR of only 8. What does this tell you?*

(A) All values are close together.

(B) Most values are clustered, but there are extreme values far from the center.

(C) The median equals the mean.

(D) The data is symmetric.

29. *A standard number cube (die) is rolled once. What is the probability of rolling a 4?*

(A) $\dfrac{4}{6}$

(B) $\dfrac{1}{4}$

(C) $\dfrac{1}{6}$

(D) $\dfrac{1}{3}$

30. *Emma earns $40 per month in allowance. Her circle graph shows how she uses it: Savings 30%, Snacks 20%, Books 15%, Charity 10%, Fun 25%. How much does Emma put toward Savings and Charity* **combined**?

(A) $10

(B) $12

(C) $16

(D) $20

Find more at
ViewMath.com/MN-Grade6

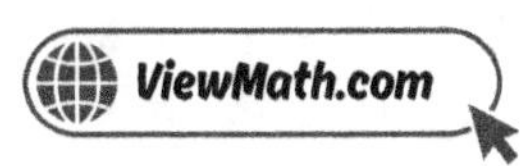

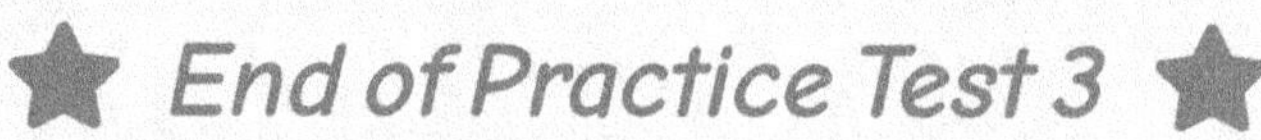

⭐ End of Practice Test 3 ⭐

Great job finishing the test!

✅ My Score

I got ___________ out of 30 questions right.

*Check your answers in the **Answer Key** at the back of the book.*

💡 *Review any questions you missed. That's how we learn!*

📊 Check Your Score Online!

*Visit **ViewMath Academy** to enter your answers and see which topics you need to review. You can also explore lessons, take quizzes, track your scores, and save your progress!*

viewmath.com/score/6.1.MN.18

Or go to viewmath.com/score and enter code: 6.1.MN.18

Practice Test 4

📋 30 Questions

✏️ Before You Start ✏️

✓ **Read each question carefully** before choosing your answer.

✓ **Show your work** on scratch paper when you need to.

✓ **Skip hard questions** and come back to them later.

✓ **Check your answers** when you're done.

✓ **Take your time** — there's no rush!

⭐ You've Got This! ⭐

Do your best and show what you know!

1. A store sells 3 pencils **for every** 1 eraser. Which ratio represents pencils to erasers?

   (A) $1:3$

   (B) $3:1$

   (C) $3:4$

   (D) $1:4$

2. Ella mixes paint in a ratio of 3 parts red to 5 parts white. She uses 24 parts of red. How many total parts of paint does she have?

   Your Answer

3. A line passes through the origin and the point $(5, 15)$. What is the ratio x to y?

   (A) $1:5$

   (B) $5:1$

   (C) $1:3$

   (D) $3:1$

4. A store is having a "Buy More, Save More" sale. Spend \$50–\$99: get 10% off. Spend \$100+: get 20% off. You want to buy items totaling \$95. Should you add a \$10 item to reach \$105? Explain.

   Your Answer

5. Which of the following is a **need** rather than a **want**?

   (A) A new video game

   (B) A pair of designer sunglasses

   (C) Groceries for the week

   (D) Concert tickets

6. The graph of a proportional relationship is always a straight line that passes through which point?

   (A) $(1, 1)$

   (B) $(0, 1)$

   (C) $(1, 0)$

   (D) $(0, 0)$

Find more at
ViewMath.com/MN-Grade6

7. If $\dfrac{a}{b} \div \dfrac{c}{d} = \dfrac{a}{b} \times \dfrac{?}{?}$, what fraction replaces the question marks?

(A) $\dfrac{d}{c}$

(B) $\dfrac{c}{d}$

(C) $\dfrac{b}{a}$

(D) $\dfrac{a}{b}$

8. Compute 3.6×0.4.

Your Answer:

9. A diver is at -40 feet and a bird is at 25 feet above sea level. Who is farther from sea level?

(A) The bird, because $25 > -40$

(B) They are the same distance from sea level

(C) The diver, because $|-40| > |25|$

(D) Neither, because you cannot compare positive and negative numbers

10. In which quadrant is the point $(-4, 7)$ located?

(A) Quadrant I

(B) Quadrant II

(C) Quadrant III

(D) Quadrant IV

11. Which phrase matches the expression $\dfrac{n}{3} + 7$?

(A) 7 more than a number n divided by 3

(B) 3 divided by n plus 7

(C) 7 times n divided by 3

(D) n plus 7, divided by 3

12. Identify the constant(s) in the expression $3m + 7 - 2m + 4$.

Your Answer:

  Find more at
ViewMath.com/MN-Grade6

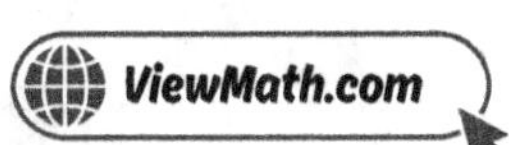

13. Look at the formula and the rectangle below. What is the perimeter of the rectangle?

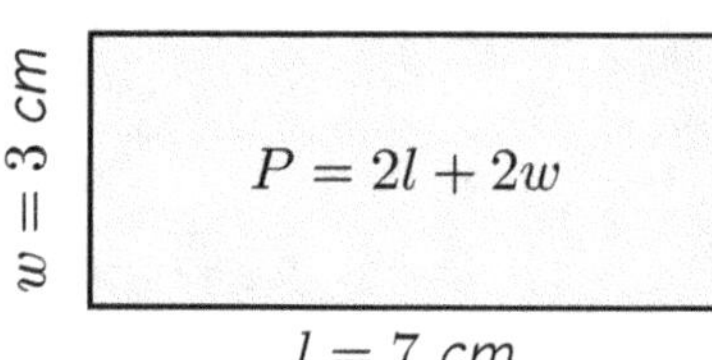

(A) 10 cm

(B) 20 cm

(C) 21 cm

(D) 42 cm

14. Simplify: $2(4m + 1) + 3m$

Your Answer

15. In the formula $P = 4s$, the variable s represents the side length of a square. What does P represent?

(A) The area of the square

(B) The perimeter of the square

(C) The number of sides

(D) The diagonal of the square

16. The thermometer shows the current temperature is 32°F. A weather report says "Temperatures will remain below freezing." Which inequality describes the predicted temperature t?

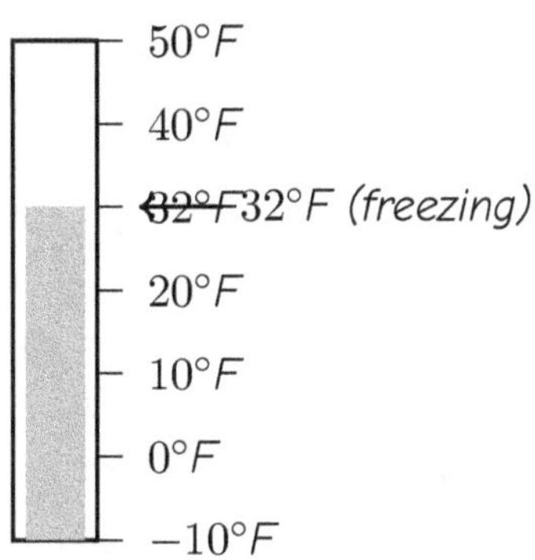

(A) $t > 32$

(B) $t \geq 32$

(C) $t < 32$

(D) $t \leq 32$

17. Which inequality has a graph where $x = -4$ is in the shaded region but $x = -2$ is NOT?

(A) $x > -3$

(B) $x < -3$

(C) $x \geq -3$

(D) $x \leq -3$

18. Using the equation $y = 3x$, complete: when $y = 21$, $x = ?$

(A) 3

(B) 7

(C) 18

(D) 63

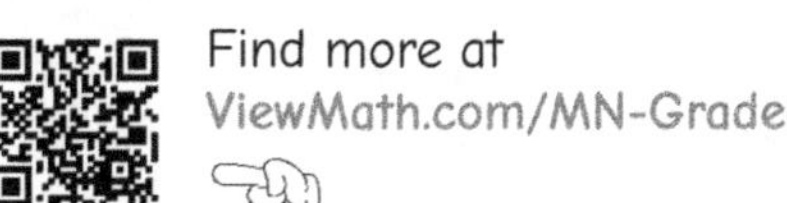
Find more at
ViewMath.com/MN-Grade6

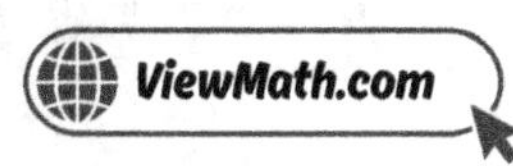

19. Two triangles both have a base of 10 cm. Triangle P has a height of 6 cm and Triangle Q has a height of 8 cm. How much greater is the area of Triangle Q?

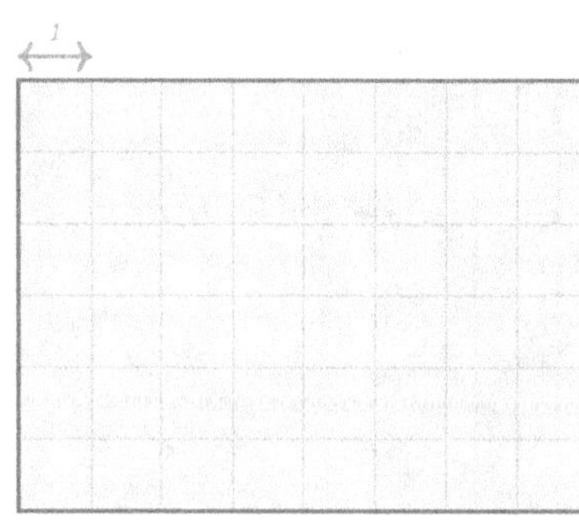

   (A) 2 cm²   (B) 10 cm²

   (C) 20 cm²   (D) 40 cm²

20. A rectangular prism has length $\frac{1}{2}$ m, width $\frac{1}{2}$ m, and height $\frac{1}{2}$ m. What is the volume?

   (A) $\frac{1}{8}$ m³   (B) $\frac{1}{4}$ m³

   (C) $\frac{3}{2}$ m³   (D) $\frac{1}{2}$ m³

21. A right triangle has vertices $(2, 1)$, $(2, 7)$, and $(8, 1)$. What is the area?

   (A) 36 square units   (B) 18 square units

   (C) 12 square units   (D) 24 square units

22. A triangular prism has 2 triangular bases and 3 rectangular faces. How many faces does it have in total?

   (A) 3   (B) 4

   (C) 5   (D) 6

23. Two circles are shown below. Circle P has a radius of 4 in and Circle Q has a radius of 8 in. The area of Circle Q is how many times the area of Circle P?

(A) 2 times   (B) 3 times

(C) 4 times   (D) 8 times

24. Explain why "What is my teacher's age?" is **not** a statistical question.

> *Your Answer:*

25. Data (in order): $4, 6, 8, 10, 12, 14, 16$. What is Q3?

(A) 10   (B) 12

(C) 14   (D) 16

26. A dot plot of temperatures shows: 68°F (1 dot), 70°F (3 dots), 72°F (4 dots), 74°F (2 dots). What is the range?

(A) 4   (B) 6

(C) 10   (D) 74

27. If the whiskers of a box plot are both short and the box is narrow, what does this tell you?

(A) The data is very spread out.

(B) The data is highly consistent with little variability.

(C) There are many outliers.

(D) The data set is very large.

28. Data: $12, 15, 18, 20, 22, 25, 28, 30$. What is the IQR?

(A) 10

(B) 11

(C) 18

(D) 8

29. Use the probability scale below. Place each event at the correct position by writing its letter on the scale. Then state each probability as a fraction or decimal.

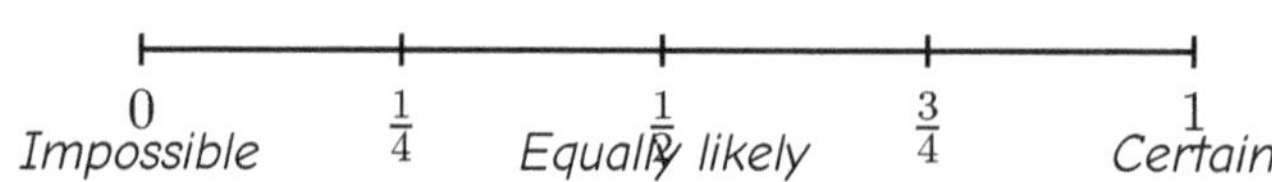

Event A: Rolling a 7 on a standard number cube.

Event B: Flipping tails on a fair coin.

Event C: Drawing a red marble from a bag that has 6 red and 2 blue marbles.

Event D: Rolling a number less than 7 on a standard number cube.

Your Answer

30. A circle graph of an election shows: Candidate A received 28%, Candidate B received 34%, and Candidate C received the rest. What percent of the votes went to Candidate C?

(A) 6%

(B) 38%

(C) 48%

(D) 62%

 End of Practice Test 4 

Great job finishing the test!

☑ My Score

I got ___________ out of 30 questions right.

Check your answers in the **Answer Key** at the back of the book.

💡 Review any questions you missed. That's how we learn!

📊 Check Your Score Online!

Visit **ViewMath Academy** to enter your answers and see which topics you need to review. You can also explore lessons, take quizzes, track your scores, and save your progress!

viewmath.com/score/6.1.MN.19

Or go to viewmath.com/score and enter code: 6.1.MN.19

5

Practice Test 5

☑ *30 Questions*

✏️ Before You Start ✏️

- ✓ **Read each question carefully** before choosing your answer.
- ✓ **Show your work** on scratch paper when you need to.
- ✓ **Skip hard questions** and come back to them later.
- ✓ **Check your answers** when you're done.
- ✓ **Take your time** — there's no rush!

⭐ You've Got This! ⭐

Do your best and show what you know!

1. A class votes on two activities: hiking and swimming. The ratio of votes for hiking to swimming is $3 : 2$. There are 25 votes total. How many voted for swimming?

Your Answer:

2. Which pair of ratios are equivalent?

(A) $2 : 3$ and $4 : 9$

(B) $2 : 3$ and $6 : 9$

(C) $2 : 3$ and $8 : 9$

(D) $2 : 3$ and $3 : 2$

3. A graph of equivalent ratios passes through $(0, 0)$ and $(3, 2)$. Which statement is true?

(A) The point $(9, 4)$ is on the line.

(B) The point $(6, 4)$ is on the line.

(C) The point $(6, 5)$ is on the line.

(D) The point $(9, 8)$ is on the line.

4. The bar chart shows the original price and sale price of three items.

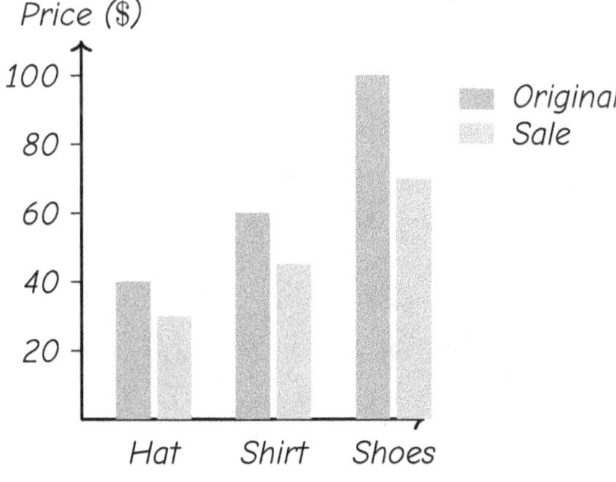

The hat was $40 and is now $30. What is the percent discount on the hat?

(A) 10%

(B) 20%

(C) 25%

(D) 30%

5. Which payment method lets you borrow money that you must pay back later, possibly with interest?

A. Cash

B. Check

C. Credit card

D. Debit card

6. Tell whether each equation is proportional or non-proportional.

(a) $y = 9x$ **(b)** $y = 4x + 6$ **(c)** $y = \dfrac{1}{2}x$

Your Answer:

7. What is $\dfrac{2}{3} \div \dfrac{4}{5}$?

A. $\dfrac{8}{15}$

B. $\dfrac{6}{5}$

C. $\dfrac{5}{6}$

D. $\dfrac{2}{15}$

8. When you multiply 0.6×0.7, how many decimal places should the product have?

A. 0

B. 1

C. 2

D. 3

9. Which statement is **true**?

A. $|-10| = -10$

B. $|-10| = 10$

C. $|-10| = 0$

D. $|-10| = -(-(-10))$

Find more at
ViewMath.com/MN-Grade6

ViewMath.com

10. Starting at the origin, you move 4 units to the left and 3 units up, then 7 units to the right and 5 units down. What point do you end at?

Your Answer

11. Write an expression for: "9 fewer than a number z."

Your Answer

12. In the expression $6(n+2)$, what are the two factors?

- (A) 6 and n
- (B) 6 and 2
- (C) n and 2
- (D) 6 and $(n+2)$

13. Evaluate $m^2 + 2m + 1$ when $m = 4$.

Your Answer

14. Simplify: $4(3y - 2)$

- (A) $12y - 2$
- (B) $12y - 8$
- (C) $7y - 2$
- (D) $7y - 6$

15. A store sells notebooks for \$3 each and pens for \$1 each. Which expression gives the total cost for n notebooks and p pens?

- (A) $3 + n + 1 + p$
- (B) $3n + p$
- (C) $4np$
- (D) $n + 3p$

Find more at
ViewMath.com/MN-Grade6

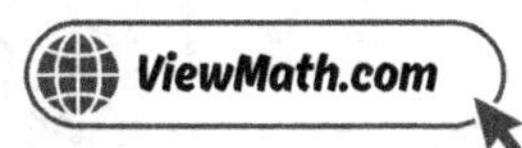

16. Is $n = 4$ a solution to $n > 4$? Is $n = 4$ a solution to $n \geq 4$? Explain both.

17. A student graphs $x > 2$ with a closed circle at 2 and shading to the right. What is the error?

(A) Should shade to the left

(B) Should use an open circle at 2

(C) Should use a circle at 0 instead

(D) There is no error

18. A plant grows 3 cm each week. Which equation relates the height h (in cm) to the number of weeks w?

(A) $h = w + 3$

(B) $h = 3w$

(C) $w = 3h$

(D) $h = w \div 3$

19. A triangular pennant has a base of 5 in and a height of 12 in. What is its area?

(A) $60\ in^2$

(B) $17\ in^2$

(C) $30\ in^2$

(D) $34\ in^2$

20. If you double the height of a rectangular prism but keep the length and width the same, what happens to the volume?

(A) The volume stays the same.

(B) The volume doubles.

(C) The volume triples.

(D) The volume quadruples.

21. An L-shaped figure has vertices $(0, 0)$, $(10, 0)$, $(10, 4)$, $(6, 4)$, $(6, 8)$, and $(0, 8)$. What is the total area?

Find more at
ViewMath.com/MN-Grade6

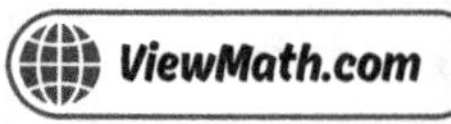

22. A rectangular prism has dimensions $10\ m$, $3\ m$, and $7\ m$. What is its surface area?

(A) $210\ m^2$

(B) $242\ m^2$

(C) $262\ m^2$

(D) $121\ m^2$

23. The circumference of a circle is $31.4\ cm$. What is the diameter of the circle? Use $\pi \approx 3.14$.

(A) $5\ cm$

(B) $10\ cm$

(C) $15.7\ cm$

(D) $20\ cm$

24. Which of the following is a statistical question?

(A) How many days are in February during a leap year?

(B) How many hours do sixth graders spend on homework each week?

(C) What is the boiling point of water in degrees Celsius?

(D) How many states are in the United States?

25. Data: $10, 10, 10, 10, 10$. What is the MAD?

(A) 0

(B) 2

(C) 5

(D) 10

26. A data set has two values that each appear 4 times, and no other value appears more than 3 times. How many modes does this data set have?

(A) 0

(B) 1

(C) 2

(D) 4

Find more at
ViewMath.com/MN-Grade6

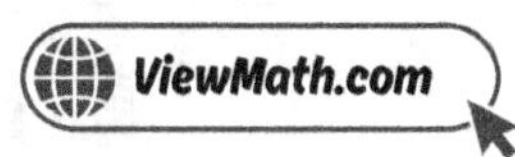

27. *A box plot shows: min $= 25$, $Q1 = 35$, median $= 50$, $Q3 = 65$, max $= 80$. Find the range and IQR.*

Your Answer:

28. *Store A daily sales: median $= \$500$, range $= \$200$. Store B daily sales: median $= \$500$, range $= \$800$. What can you conclude?*

(A) Both stores are equally consistent.

(B) Store A is more consistent in its daily sales.

(C) Store B is more consistent in its daily sales.

(D) Store B has lower typical sales.

29. *The spinner below has 8 equal sections. Find each probability as a fraction in simplest form.*
(a) $P(red)$ (b) $P(blue)$ (c) $P(not\ green)$

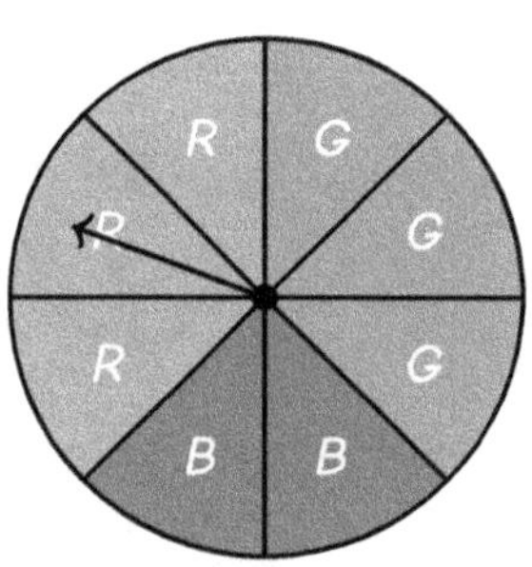

Your Answer:

30. *A school has 600 students. A circle graph shows: Grade 6 is 35%, Grade 7 is 40%, Grade 8 is 25%. How many more students are in Grade 7 than in Grade 8?*

(A) 15

(B) 40

(C) 90

(D) 150

Find more at
ViewMath.com/MN-Grade6

ViewMath.com

 # End of Practice Test 5 

Great job finishing the test!

 My Score

I got _____________ out of 30 questions right.

*Check your answers in the **Answer Key** at the back of the book.*

💡 *Review any questions you missed. That's how we learn!*

📊 Check Your Score Online!

Visit **ViewMath Academy** to enter your answers and see which topics you need to review. You can also explore lessons, take quizzes, track your scores, and save your progress!

viewmath.com/score/6.1.MN.20

Or go to viewmath.com/score and enter code: 6.1.MN.20

6

Practice Test 6

 30 Questions

✏ Before You Start ✏

- ✓ **Read each question carefully** before choosing your answer.
- ✓ **Show your work** on scratch paper when you need to.
- ✓ **Skip hard questions** and come back to them later.
- ✓ **Check your answers** when you're done.
- ✓ **Take your time** — there's no rush!

 ⭐ You've Got This! ⭐

Do your best and show what you know!

1. A smoothie recipe uses 3 bananas **for every** 4 cups of yogurt. Write the ratio of bananas to yogurt. Then write the ratio of yogurt to bananas.

Your Answer

2. Look at this ratio table. What is the missing value?

Apples	Oranges
3	5
6	?

(A) 8

(B) 10

(C) 15

(D) 11

3. Which table matches a graph that passes through $(0,0)$, $(2,3)$, and $(4,6)$?

(A)

x	y
2	3
6	9

(B)

x	y
2	3
6	8

(C)

x	y
3	2
6	9

(D)

x	y
2	4
6	12

4. A backpack is $60 with a 25% off coupon. What is the sale price?

(A) $35

(B) $40

(C) $45

(D) $50

Find more at
ViewMath.com/MN-Grade6

5. Which is the best reason to save money regularly?

- (A) So you can spend more on wants right away
- (B) So you have money for unexpected expenses or future goals
- (C) So the bank will give you a credit card
- (D) So you never have to earn income again

6. In a proportional relationship, $y = kx$. What is k called?

- (A) The slope intercept
- (B) The variable
- (C) The constant of proportionality
- (D) The y-intercept

7. A recipe uses $\frac{2}{3}$ cup of sugar per batch. You have 4 cups of sugar. How many batches can you make?

- (A) $\frac{8}{3}$
- (B) 2
- (C) 6
- (D) 12

8. What is $9.36 \div 1.2$?

- (A) 78
- (B) 0.78
- (C) 7.8
- (D) 7.08

9. What is the opposite of -3?

- (A) -3
- (B) 0
- (C) $\frac{1}{3}$
- (D) 3

Find more at
ViewMath.com/MN-Grade6

ViewMath.com

10. *Starting at the origin, how do you plot the point $(-3, 4)$?*

(A) Move 3 units right and 4 units up

(B) Move 3 units left and 4 units down

(C) Move 3 units left and 4 units up

(D) Move 4 units left and 3 units up

11. *Which expression represents "6 more than half of a number n"?*

(A) $6 + 2n$

(B) $\dfrac{n+6}{2}$

(C) $\dfrac{n}{2} + 6$

(D) $n + 3$

12. *Look at the expression map below. Which label correctly identifies Part C?*

$$\underbrace{5}_{A}\ \underbrace{x}_{B} + \underbrace{3}_{C}$$

(A) *A variable.*

(B) *A coefficient*

(C) *A constant*

(D) *A factor*

13. *Evaluate $\dfrac{x^2}{4}$ when $x = 8$.*

Your Answer

14. *Simplify:* $3(x + 4) + 2x$

(A) $5x + 4$

(B) $5x + 12$

(C) $6x + 12$

(D) $3x + 6$

15. Emma earns $8 per hour babysitting. She also got a $15 tip. Which expression gives her total earnings for h hours?

  (A) $8 + 15h$ (B) $23h$

  (C) $8h + 15$ (D) $8h - 15$

16. Which inequality represents "you need more than $25 to buy the video game"?

  (A) $d \leq 25$ (B) $d < 25$

  (C) $d > 25$ (D) $d \geq 25$

17. Write the inequality that matches this description: closed circle at -4, shade to the left.

Your Answer:

18. A school bus holds 50 students. How many buses b are needed for s students? Write an equation.

Your Answer:

19. A triangular garden has a base of 12 ft and a height of 9 ft. How many square feet of soil are needed to cover it?

  (A) $108 \ ft^2$ (B) $54 \ ft^2$

  (C) $21 \ ft^2$ (D) $42 \ ft^2$

20. A rectangular prism is made entirely of 1-cm cubes. It is 4 cubes long, 3 cubes wide, and 5 cubes tall. What is the volume?

Your Answer:

Find more at
ViewMath.com/MN-Grade6

ViewMath.com

21. A rectangle on the coordinate plane has an area of 56 square units. Its length is 8 units. What is the width?

(A) 6 units

(B) 7 units

(C) 8 units

(D) 48 units

22. A box is 15 in long, 6 in wide, and 4 in tall. How many square inches of cardboard are needed to make the box?

Your Answer:

23. What is the area of a circle with radius 9 m? Use $\pi \approx 3.14$.

(A) $28.26 \ m^2$

(B) $56.52 \ m^2$

(C) $254.34 \ m^2$

(D) $508.68 \ m^2$

24. Which of the following is a statistical question?

(A) What year was the school built?

(B) What is the name of our principal?

(C) How many minutes does each student exercise per day?

(D) How many continents are there?

25. Data: $20, 22, 24, 26, 28$. The mean is 24. What is the MAD?

(A) 2

(B) 2.4

(C) 4

(D) 8

Find more at
ViewMath.com/MN-Grade6

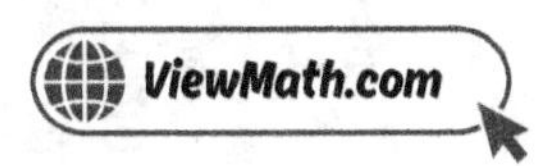

26. Which display is best for showing favorite colors of students?

(A) Histogram

(B) Dot plot

(C) Frequency table or bar graph

(D) Box plot

27. Which data set has the five-number summary: $5, 10, 15, 20, 25$?

(A) $5, 10, 15, 20, 25$

(B) $5, 8, 10, 15, 18, 20, 22, 25$

(C) $5, 7, 10, 12, 15, 18, 20, 23, 25$

(D) Any of these could have that five-number summary.

28. Daily steps for Week 1: median $= 8,000$, IQR $= 2,000$. Week 2: median $= 9,500$, IQR $= 1,200$. Which week shows a higher typical step count and more consistent activity?

(A) Week 1 for both

(B) Week 2 for both

(C) Week 1 higher, Week 2 more consistent

(D) Week 2 higher, Week 1 more consistent

29. A fair coin is flipped once. What is the probability of landing on heads?

(A) $\dfrac{1}{4}$

(B) $\dfrac{1}{3}$

(C) $\dfrac{1}{2}$

(D) 1

30. A circle graph shows how 200 students get to school: Bus 40%, Car 25%, Walk 20%, Bike 15%. How many students take the bus?

(A) 30

(B) 60

(C) 80

(D) 100

 *End of Practice Test 6* 

Great job finishing the test!

📋 *My Score*

I got _____________ out of 30 questions right.

📊 *Check Your Score Online!*

*Visit **ViewMath Academy** to enter your answers and see which topics you need to review. You can also explore lessons, take quizzes, track your scores, and save your progress!*

viewmath.com/score/6.1.MN.21

Practice Test 7

☑ *30 Questions*

✏ Before You Start ✏

- ✓ **Read each question carefully** before choosing your answer.
- ✓ **Show your work** on scratch paper when you need to.
- ✓ **Skip hard questions** and come back to them later.
- ✓ **Check your answers** when you're done.
- ✓ **Take your time** — there's no rush!

⭐ You've Got This! ⭐

Do your best and show what you know!

1. A lemonade stand sells 7 cups of lemonade **for each** 2 cups of iced tea. If they sold 21 cups of lemonade, how many cups of iced tea did they sell?

Your Answer:

2. A ratio table for pens to notebooks starts with $2 : 3$. Which of these is a valid row?

(A) $4 : 5$

(B) $8 : 12$

(C) $6 : 12$

(D) $10 : 12$

3. A ratio graph passes through $(4, 10)$ and the origin. List two other points on this line.

Your Answer:

4. What is 100% of any number?

(A) 0

(B) Half the number

(C) The number itself

(D) Double the number

5. Explain one advantage and one disadvantage of using a credit card instead of cash.

Your Answer:

6. A relationship is **proportional** if the ratio $\frac{y}{x}$ is the same for every pair of values. Which table shows a proportional relationship?

(A) x: $1, 2, 3$ y: $4, 7, 10$

(B) x: $1, 2, 3$ y: $5, 10, 15$

(C) x: $1, 2, 3$ y: $3, 5, 7$

(D) x: $1, 2, 3$ y: $2, 5, 8$

Find more at
ViewMath.com/MN-Grade6

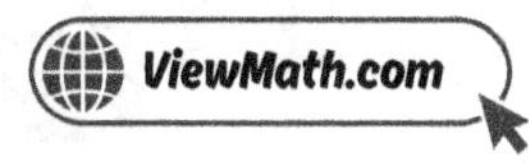

7. A carpenter has $2\frac{1}{2}$ feet of wood. Each shelf needs $\frac{5}{8}$ of a foot. How many shelves can the carpenter cut?

Your Answer:

8. Sam buys 3 folders that each cost $2.75. How much does he spend in all?

(A) $8.75

(B) $8.25

(C) $6.25

(D) $5.75

9. What is the opposite of 5?

(A) 5

(B) -5

(C) 0

(D) $\frac{1}{5}$

10. Where is the point $(-6, 0)$ located on the coordinate plane?

(A) On the y-axis

(B) On the x-axis

(C) In Quadrant II

(D) In Quadrant III

11. Write a word phrase for the expression $5x + 3$.

Your Answer:

12. Which expression has exactly 4 terms?

(A) $2x + 3y$

(B) $a + b + c$

(C) $5m - 2n + 7 + m$

(D) $4p$

Find more at
ViewMath.com/MN-Grade6

ViewMath.com

13. *The perimeter of a rectangle is $P = 2l + 2w$. What is P when $l = 9$ and $w = 4$?*

(A) 13

(B) 22

(C) 26

(D) 36

14. *Simplify:* $5p + 3 + 2p + 4p - 1$

(A) $11p + 2$

(B) $7p + 2$

(C) $11p + 4$

(D) $12p$

15. *A taxi charges \$3 plus \$2 per mile. Write an expression for the cost of a ride that is m miles long.*

Your Answer:

16. *A bag can hold at most 50 pounds. Which inequality represents the allowed weight w?*

(A) $w > 50$

(B) $w < 50$

(C) $w \le 50$

(D) $w \ge 50$

17. *When graphing $x > 4$ on a number line, what kind of circle goes at 4?*

(A) Closed circle (filled in)

(B) Open circle (not filled in)

(C) No circle — just an arrow

(D) A square

18. *On a graph of $y = 4x$, which ordered pair is on the line?*

(A) $(2, 6)$

(B) $(3, 12)$

(C) $(4, 12)$

(D) $(5, 25)$

Find more at
ViewMath.com/MN-Grade6

19. Which of the following is true about the height of a triangle?

    (A) The height is always a side of the triangle.   (B) The height must be perpendicular to the base.

    (C) The height is always the longest measurement.   (D) The height must equal the base.

20. A box is 2.5 cm long, 4 cm wide, and 6 cm tall. What is the volume?

Your Answer:

21. An L-shaped figure has vertices $(0,0)$, $(8,0)$, $(8,3)$, $(4,3)$, $(4,7)$, and $(0,7)$. A student splits it into two rectangles. Which pair of rectangles correctly covers the figure?

    (A) 8×7 and 4×3 (B) 8×3 and 4×4

    (C) 4×7 and 4×3 (D) 8×3 and 4×7

22. A rectangular prism has dimensions 4 cm by 4 cm by 9 cm. What is the surface area?

    (A) $144\ cm^2$ (B) $176\ cm^2$

    (C) $160\ cm^2$ (D) $208\ cm^2$

23. A circle has a diameter of 15 cm. Find the circumference of the circle. Use $\pi \approx 3.14$.

Your Answer:

24. A student surveyed classmates about how many books they read last month and made the dot plot below.

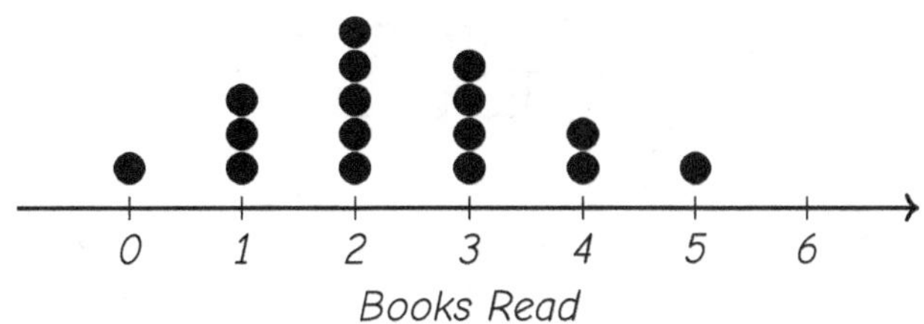

What does this dot plot confirm about the survey question "How many books did you read last month?"

(A) It is not statistical because only whole numbers appear.

(B) It is statistical because the answers vary from 0 to 5.

(C) It is not statistical because some students read the same number.

(D) It is statistical because there are exactly 16 dots.

25. Data: $100, 105, 110, 115, 120, 125, 130$. Find the IQR.

Your Answer

26. A data set is: $5, 5, 5, 5, 5$. What is the mode?

(A) 0

(B) 1

(C) 5

(D) There is no mode.

27. The two box plots below represent daily sales (in dollars) at two stores over the same month.

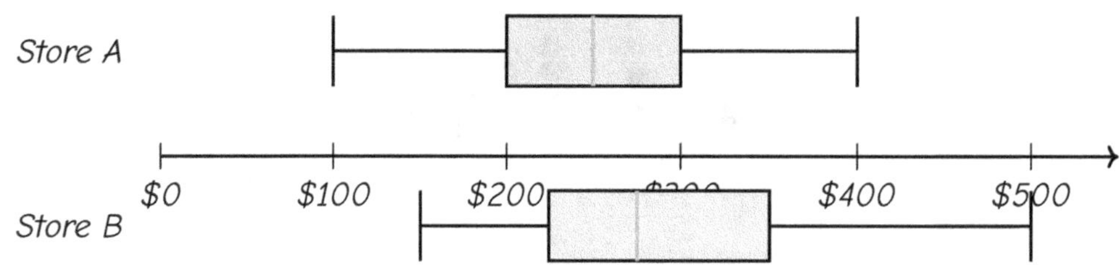

Compare the two stores. Which store has higher typical sales? Which has more predictable sales? Use the box plots to support your answers.

Your Answer.

28. Which two measures together best describe the "center" and "spread" of a data set?

(A) Mean and mode

(B) Median and IQR

(C) Range and maximum

(D) Minimum and maximum

29. Event A has a probability of $\frac{2}{5}$ and Event B has a probability of $\frac{3}{4}$. Which statement is true?

(A) Event A is more likely than Event B

(B) Event B is more likely than Event A

(C) Both events are equally likely

(D) Neither event can happen

30. A sector in a circle graph represents 75% of the data. Which fraction in simplest form equals 75%?

(A) $\frac{1}{4}$

(B) $\frac{3}{5}$

(C) $\frac{3}{4}$

(D) $\frac{7}{10}$

Find more at
ViewMath.com/MN-Grade6

ViewMath.com

 # End of Practice Test 7 

Great job finishing the test!

 My Score

I got _____________ out of 30 questions right.

*Check your answers in the **Answer Key** at the back of the book.*

💡 *Review any questions you missed. That's how we learn!*

📊 Check Your Score Online!

Visit **ViewMath Academy** to enter your answers and see which topics you need to review. You can also explore lessons, take quizzes, track your scores, and save your progress!

viewmath.com/score/6.1.MN.22

Or go to viewmath.com/score and enter code: 6.1.MN.22

Practice Test 8

☑ *30 Questions*

✏️ Before You Start ✏️

- ✔ **Read each question carefully** before choosing your answer.
- ✔ **Show your work** on scratch paper when you need to.
- ✔ **Skip hard questions** and come back to them later.
- ✔ **Check your answers** when you're done.
- ✔ **Take your time** — there's no rush!

⭐ You've Got This! ⭐

Do your best and show what you know!

1. The ratio of cats to dogs at a pet store is $3 : 4$. Each part in the tape diagram represents 2 animals. How many dogs are there?

(A) 6

(B) 8

(C) 3

(D) 4

2. The graph below shows points that represent equivalent ratios of flour to sugar in a recipe.

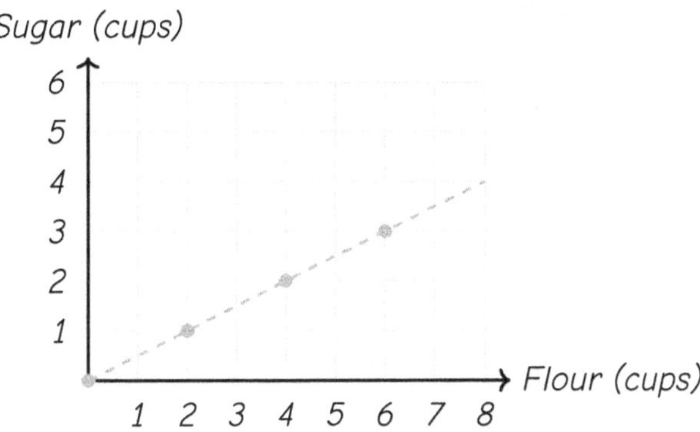

Part A: What is the ratio of flour to sugar?

Part B: If you need 8 cups of flour, how many cups of sugar do you need?

Your Answer:

3. True or false: a graph of equivalent ratios can be a curved line. Explain.

Your Answer:

4. Which represents 35% as a fraction in simplest form?

(A) $\dfrac{35}{10}$

(B) $\dfrac{7}{20}$

(C) $\dfrac{35}{50}$

(D) $\dfrac{7}{10}$

5. Nina puts \$800 in a savings account that earns 6% simple interest per year. How much interest does she earn after 2 years, and what is the total amount in her account?

Your Answer:

6. The coordinate plane below shows points from two different relationships.

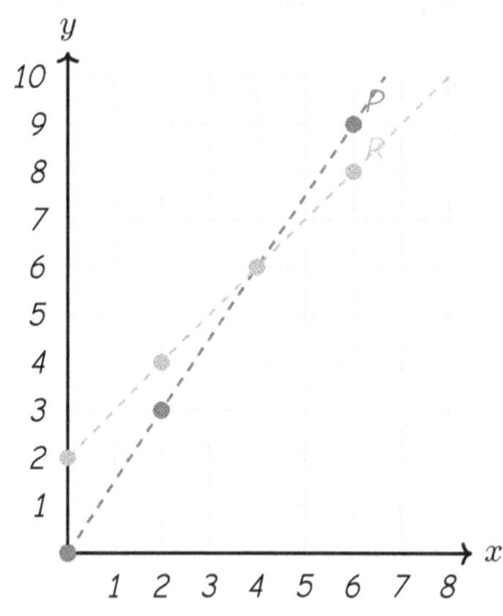

Which statement is true?

(A) Both P and R are proportional.

(B) P is non-proportional; R is proportional.

(C) P is proportional; R is non-proportional.

(D) Neither P nor R is proportional.

7. Kai says $\frac{2}{7} \div \frac{5}{6} = \frac{12}{35}$. Is Kai correct?

(A) No, the answer is $\frac{10}{42}$

(B) Yes, $\frac{12}{35}$ is correct

(C) No, the answer is $\frac{35}{12}$

(D) No, the answer is $\frac{7}{15}$

8. *What is* 0.12×0.5?

(A) 0.6 (B) 0.06

(C) 0.006 (D) 6.0

9. *What is the opposite of* 0?

(A) 1 (B) -1

(C) 0 (D) *There is no opposite of* 0

10. **Part A:** *In which quadrant is the point* $(-8, 1)$ *located?*

Part B: *If* $(-8, 1)$ *is reflected across the* y-*axis, what are the coordinates of its image and in which quadrant does it lie?*

Part C: *If the original point* $(-8, 1)$ *is instead reflected across the* x-*axis, what are the coordinates of its image and in which quadrant does it lie?*

Your Answer

11. *The table below shows the relationship between a word phrase and an expression. Which expression belongs in the blank?*

Word Phrase	Expression
A number plus 4	$n + 4$
3 times a number	$3n$
8 less than a number	?
A number divided by 2	$n \div 2$

(A) $8 - n$ (B) $n - 8$

(C) $8n$ (D) $n + 8$

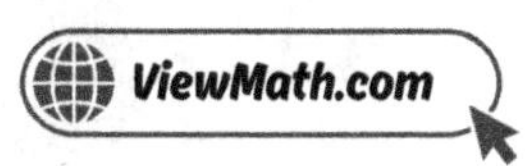

12. What is the coefficient of d in the expression $15 - d + 3$?

(A) 0

(B) 1

(C) -1

(D) 15

13. A parking garage charges \$5 plus \$3 per hour. How much does it cost to park for 6 hours? Use the expression $5 + 3h$.

Your Answer:

14. The rectangle below has the dimensions shown. Write a simplified expression for its perimeter.

$$2x + 5$$

$x + 1$ $\qquad$ $x + 1$

$$2x + 5$$

Your Answer:

15. Marcus has n baseball cards. He buys 12 more and then gives 5 to his friend. Write an expression for how many cards he has now.

Your Answer:

16. A sign says "No more than 4 guests per room." Write an inequality for the number of guests g.

Your Answer:

Find more at
ViewMath.com/MN-Grade6

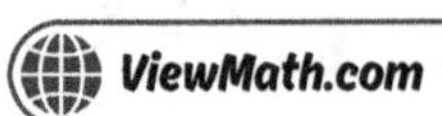

17. Is $x = 3$ a solution to $x \leq 3$? Is $x = 3$ a solution to $x < 3$? Explain.

Your Answer:

18. In the equation $y = x + 4$, which statement is true?

(A) x is the dependent variable

(B) y is always 4

(C) x is the independent variable

(D) y decreases as x increases

19. A triangular flower bed has a base of 4.5 m and a height of 6 m. What is its area?

Your Answer:

20. Two boxes are shown below with their dimensions. Which box has the greater volume?

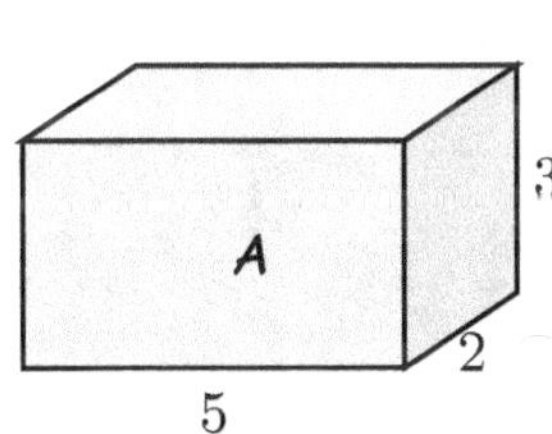

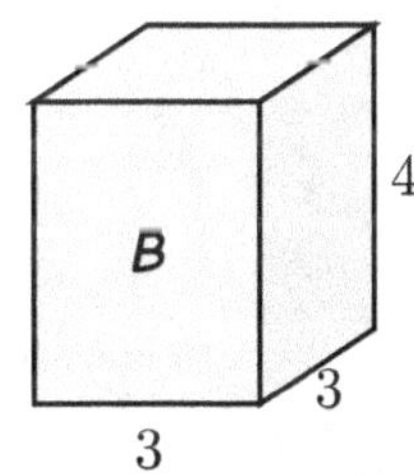

(A) Box A $(V = 30)$

(B) Box B $(V = 36)$

(C) They have the same volume.

(D) Box B $(V = 30)$

21. A right triangle has vertices $(-3, -2)$, $(5, -2)$, and $(-3, 4)$. What is the area?

Your Answer:

Find more at
ViewMath.com/MN-Grade6

22. A cube has edges of 5 in. What is the surface area?

(A) $25\ in^2$

(B) $125\ in^2$

(C) $150\ in^2$

(D) $75\ in^2$

23. A circle has a radius of 7 cm. What is its circumference? Use $\pi \approx 3.14$

(A) $21.98\ cm$

(B) $43.96\ cm$

(C) $87.92\ cm$

(D) $153.86\ cm$

24. Write a statistical question you could ask your classmates about food.

Your Answer:

25. Data: $50, 55, 60, 65, 70$. The mean is 60. Find the MAD.

Your Answer:

26. You survey 15 classmates about their favorite fruit. Which display should you start with?

(A) Histogram, because it groups data into intervals.

(B) Frequency table, because you need to count each category first.

(C) Box plot, because you need the five-number summary.

(D) Dot plot, because it shows individual numerical values.

27. A box plot has: min $= 60$, Q1 $= 70$, median $= 75$, Q3 $= 85$, max $= 100$. What percent of data falls between 70 and 85?

Your Answer:

Find more at
ViewMath.com/MN-Grade6

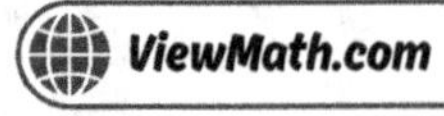

28. Which pair of measures should you report when outliers are present?

(A) Mean and range

(B) Mode and maximum

(C) Median and IQR

(D) Mean and IQR

29. A bag contains 4 green marbles, 6 red marbles, and 10 blue marbles. What is the probability of randomly drawing a green marble? Write your answer as a percent.

Your Answer

30. A circle graph shows that 15% of all donations went to animal shelters. If animal shelters received $450, what was the total amount of donations?

(A) $1,500

(B) $3,000

(C) $4,500

(D) $30,000

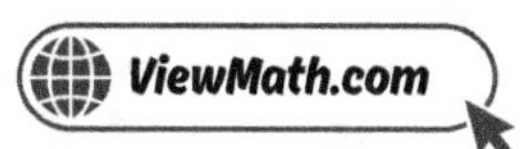

 # ★ End of Practice Test 8 ★ 

Great job finishing the test!

 ## ☑ My Score

I got _____________ out of 30 questions right.

*Check your answers in the **Answer Key** at the back of the book.*

💡 *Review any questions you missed. That's how we learn!*

📊 Check Your Score Online!

*Visit **ViewMath Academy** to enter your answers and see which topics you need to review. You can also explore lessons, take quizzes, track your scores, and save your progress!*

viewmath.com/score/6.1.MN.23

Or go to viewmath.com/score and enter code: 6.1.MN.23

9

Practice Test 9

 30 Questions

 Before You Start

- ✓ **Read each question carefully** before choosing your answer.
- ✓ **Show your work** on scratch paper when you need to.
- ✓ **Skip hard questions** and come back to them later.
- ✓ **Check your answers** when you're done.
- ✓ **Take your time** — there's no rush!

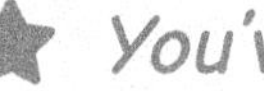 You've Got This! 

Do your best and show what you know!

1. The ratio of red to blue to green beads is $1 : 3 : 2$. There are 18 beads total. How many blue beads are there?

   (A) 3

   (B) 6

   (C) 9

   (D) 12

2. The ratio of red to blue beads is $1 : 4$. If there are 3 red beads, how many blue beads are there?

   (A) 4

   (B) 7

   (C) 12

   (D) 8

3. Two ratio graphs both pass through the origin. Line A goes through $(1, 4)$ and Line B goes through $(1, 3)$. Which line is steeper?

   (A) Line A

   (B) Line B

   (C) They have the same steepness.

   (D) Cannot be determined.

4. A video game costs $50. Sales tax is 6%. What is the total cost?

   (A) $53

   (B) $56

   (C) $55

   (D) $52

5. Tyler earns $80 from a weekend job. He spends $50 and saves the rest. What percent of his earnings does he save?

   (A) 25%

   (B) 30%

   (C) 37.5%

   (D) 62.5%

Find more at
ViewMath.com/MN-Grade6

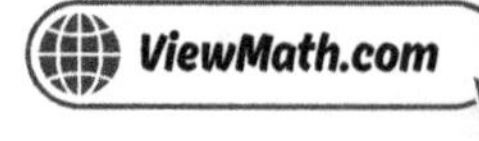

6. A table shows the values $(2, 8)$, $(3, 12)$, and $(5, 20)$. What is the constant of proportionality k?

(A) 2

(B) 3

(C) 4

(D) 6

7. Each serving of juice is $\frac{2}{5}$ cup. A jug holds $\frac{4}{5}$ cup. How many servings are in the jug?

(A) $\frac{8}{25}$

(B) $\frac{2}{5}$

(C) $\frac{5}{2}$

(D) 2

8. The table shows the distances a cyclist rode each day.

Day	Distance (km)
Monday	8.76
Tuesday	12.30
Wednesday	6.48

Part A: What is the total distance the cyclist rode over the three days?

Part B: If the cyclist wants to divide the total distance equally over the 3 days, how many kilometers per day is that?

Your Answer:

9. A number plus its opposite always equals which value?

(A) 1

(B) The number itself

(C) -1

(D) 0

10. *A triangle is drawn on the coordinate plane below with vertices at A, B, and C.*

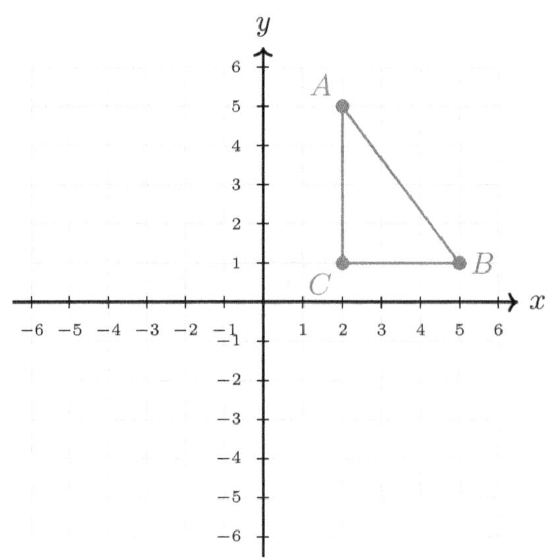

Part A: *Write the coordinates of each vertex: A, B, and C.*

Part B: *If the triangle is reflected across the y-axis, write the coordinates of the new vertices A', B', and C'.*

Part C: *In which quadrant will the reflected triangle be located?*

Your Answer:

11. *Which expression represents "the sum of 9 and the product of 4 and t"?*

(A) $9 + 4 + t$

(B) $9 + 4t$

(C) $9(4 + t)$

(D) $9 \times 4t$

12. *How many terms are in the expression $4x + 9 - 2y$?*

(A) 2

(B) 3

(C) 4

(D) 5

13. Evaluate $6x - x^2$ when $x = 4$.

(A) 8

(B) 20

(C) -8

(D) 40

14. Simplify: $7n + 4n$

(A) $11n$

(B) $28n$

(C) $11n^2$

(D) $74n$

15. In $P = 4s$, can s be any positive number, or is it one specific number?

(A) Only $s = 1$

(B) Only whole numbers

(C) Any positive number — the formula works for all squares

(D) Only $s = 4$

16. Which of the following is a solution to $n \leq 8$?

(A) $n = 9$

(B) $n = 8.5$

(C) $n = 8$

(D) $n = 10$

17. Which value is NOT in the solution set of $x < 10$?

(A) $x = 0$

(B) $x = 9.9$

(C) $x = -5$

(D) $x = 10$

Find more at
ViewMath.com/MN-Grade6

18. Using $d = 60t$, what is the distance after 3 hours?

   (A) 20 miles

   (B) 63 miles

   (C) 180 miles

   (D) 360 miles

19. A triangle has base 20 in and height 7 in. What is the area?

   (A) $140\ in^2$

   (B) $27\ in^2$

   (C) $70\ in^2$

   (D) $54\ in^2$

20. A box has a volume of $72\ in^3$. The base is 6 in by 4 in. What is the height?

   (A) 2 in

   (B) 3 in

   (C) 6 in

   (D) 12 in

21. A right triangle has vertices $(1, 1)$, $(1, 9)$, and $(5, 1)$. What is the area?

   (A) 32 square units

   (B) 16 square units

   (C) 20 square units

   (D) 8 square units

22. What is the surface area of a rectangular prism with length 6 cm, width 4 cm, and height 3 cm?

   (A) $72\ cm^2$

   (B) $108\ cm^2$

   (C) $54\ cm^2$

   (D) $96\ cm^2$

23. What is the area of a semicircle with a radius of 10 cm? Use $\pi \approx 3.14$

   (A) $31.4\ cm^2$

   (B) $157\ cm^2$

   (C) $314\ cm^2$

   (D) $628\ cm^2$

Find more at
ViewMath.com/MN-Grade6

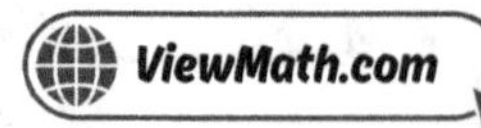

24. Which question is **not** statistical?

(A) How many siblings do students in our school have?

(B) How far do students travel to get to school?

(C) What is 12×9?

(D) How many pets do families in our neighborhood own?

25. Look at the two dot plots below.

Data Set A

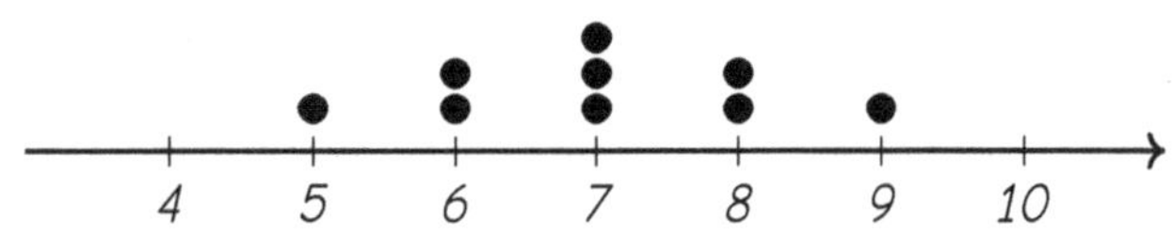

Data Set B

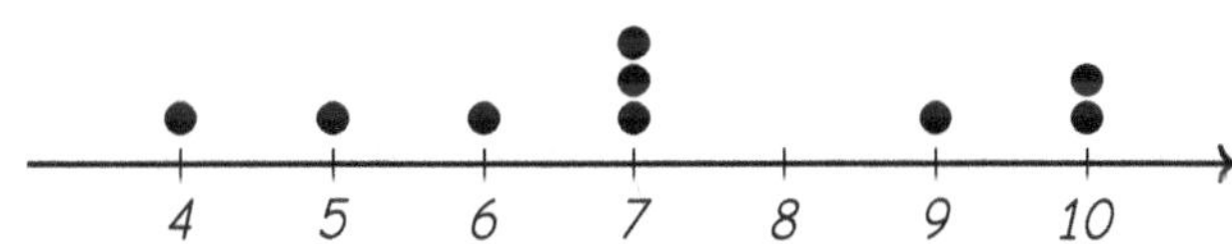

Which data set has a larger range?

(A) Data Set A

(B) Data Set B

(C) They have the same range.

(D) Cannot be determined.

26. You want to know whether to display data as a dot plot, histogram, or frequency table. Which question should you ask first?

(A) What is the mean of the data?

(B) How many data points are there, and is the data numerical or categorical?

(C) What color should the bars be?

(D) Is the data symmetric or skewed?

27. Look at the two box plots below showing test scores for two classes.

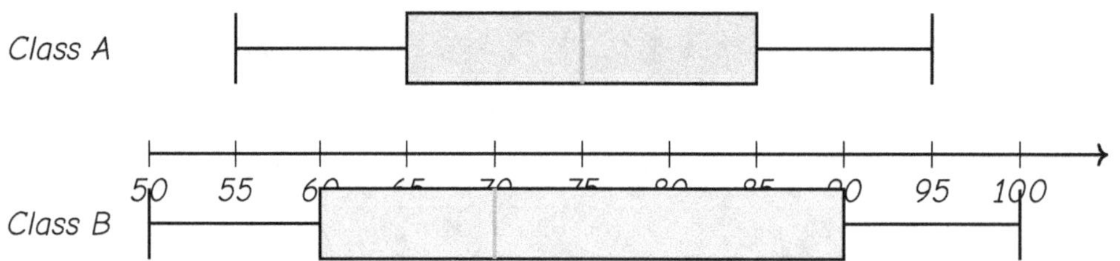

Which class has more consistent scores?

(A) Class A, because its box is narrower.

(B) Class B, because its box is wider.

(C) Both are equally consistent.

(D) Class B, because its median is higher.

28. Which measure of spread is affected by every single data value?

(A) IQR

(B) Range

(C) Median

(D) Mode

29. A jar holds 10 jellybeans: 4 cherry, 4 grape, and 2 lemon. Which event is **equally likely** as drawing a cherry jellybean?

(A) Drawing a lemon jellybean

(B) Drawing a grape jellybean

(C) Not drawing a jellybean

(D) Drawing a cherry or lemon jellybean

30. In a circle graph of 300 students, 40% chose basketball, 35% chose soccer, and the rest chose swimming. How many students chose swimming?

Your Answer:

Find more at
ViewMath.com/MN-Grade6

ViewMath.com

 # End of Practice Test 9 

Great job finishing the test!

 My Score

I got ___________ out of 30 questions right.

Check your answers in the **Answer Key** at the back of the book.

💡 Review any questions you missed. That's how we learn!

📊 Check Your Score Online!

Visit **ViewMath Academy** to enter your answers and see which topics you need to review. You can also explore lessons, take quizzes, track your scores, and save your progress!

viewmath.com/score/6.1.MN.24

Or go to viewmath.com/score and enter code: 6.1.MN.24

10

Practice Test 10

30 Questions

✏️ Before You Start ✏️

- ✓ **Read each question carefully** before choosing your answer.
- ✓ **Show your work** on scratch paper when you need to.
- ✓ **Skip hard questions** and come back to them later.
- ✓ **Check your answers** when you're done.
- ✓ **Take your time** — there's no rush!

⭐ You've Got This! ⭐

Do your best and show what you know!

1. The tape diagram below shows the ratio of apple juice to orange juice in a drink.

Apple: ☐☐☐
Orange: ☐☐☐☐☐

If each part equals 4 ounces, how many total ounces of drink are there?

(A) 12

(B) 20

(C) 32

(D) 8

2. The ratio table below shows the relationship between bags of seed and the area of garden they cover.

Bags of Seed	Area (sq ft)
2	50
4	100
6	?
8	200

What is the missing value?

(A) 125

(B) 150

(C) 175

(D) 160

3. Juice costs \$2 per bottle. Which point would **not** appear on the graph of this ratio?

(A) $(1, 2)$

(B) $(3, 6)$

(C) $(4, 10)$

(D) $(5, 10)$

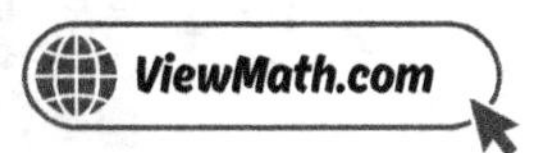

4. A sweater costs $45. You have a 20% off coupon, but must pay 8% sales tax on the sale price. What is the total cost?

(A) $36.00

(B) $38.88

(C) $40.50

(D) $42.12

5. Marcus has $150 on a credit card at 10% annual interest and $150 in a savings account earning 2% annual interest. After one year with no payments or withdrawals, what is the difference between the interest he **owes** and the interest he **earns**?

(A) $8

(B) $12

(C) $15

(D) $18

6. Two students track the distance they walk. Their data is graphed below.

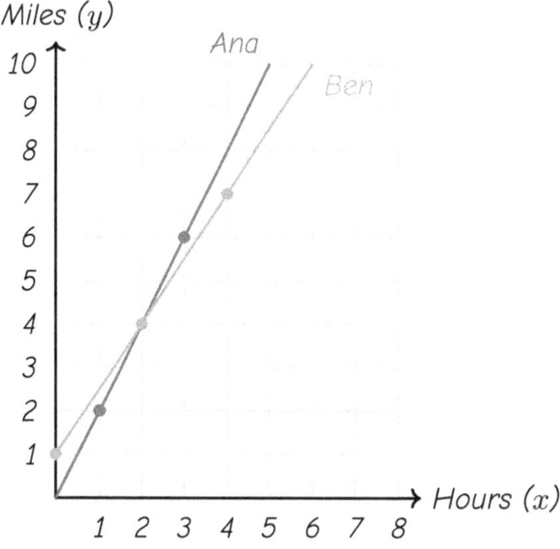

Part A: Which student's data shows a proportional relationship? Write the equation for that student.
Part B: Explain why the other student's data is non-proportional using the graph.

Your Answer:

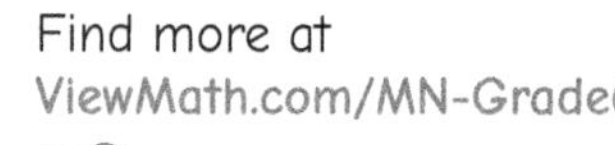
Find more at
ViewMath.com/MN-Grade6

7. Evaluate: $\dfrac{7}{10} \div \dfrac{2}{5}$

Your Answer:

8. Compute $8.64 \div 0.4$.

Your Answer:

9. What is $|0|$?

(A) There is no answer

(B) -1

(C) 1

(D) 0

10. In which quadrant is the point $(2, -5)$ located?

(A) Quadrant I

(B) Quadrant II

(C) Quadrant III

(D) Quadrant IV

11. Look at the balance model below. Each triangle represents the same unknown number x, and each small square represents 1. Write an expression for the total value shown on the balance.

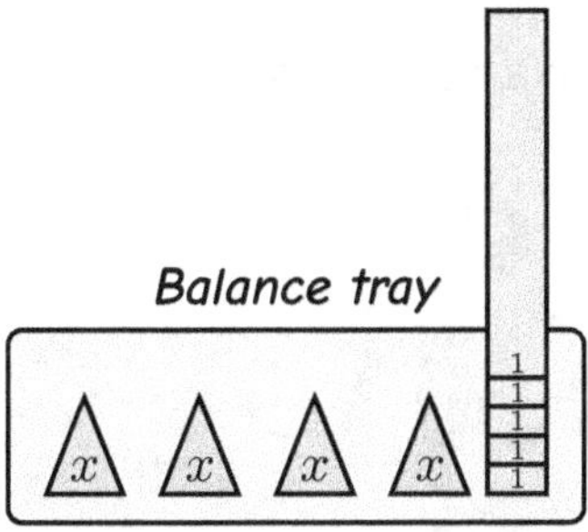

Your Answer:

Find more at
ViewMath.com/MN-Grade6

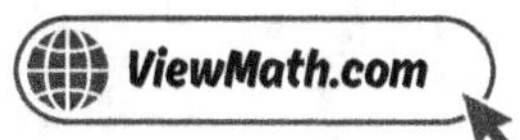

12. In the term $\dfrac{n}{5}$, what are the factors?

   (A) n and 5   (B) n and $\dfrac{1}{5}$

   (C) 1 and n (D) 5 and $\dfrac{1}{n}$

13. The table below shows the cost of renting a bike using the expression $C = 8 + 3h$, where h is the number of hours. Fill in the missing values A and B.

Hours (h)	Cost (C)
1	$11
2	A
3	$17
5	B

Your Answer

14. Which expression is equivalent to $7(k - 3) + 5$?

   (A) $7k - 16$ (B) $7k + 2$

   (C) $7k - 26$ (D) $7k - 21$

15. You have 100 pages to read and read r pages each day. Which expression tells how many pages are left after 4 days?

   (A) $100 + 4r$ (B) $100 - 4r$

   (C) $4r - 100$ (D) $100r - 4$

16. *Write an inequality: You must be no fewer than 16 years old to drive. Let a = age.*

Your Answer:

17. *Describe the graph of $n \leq 5$.*

(A) *Open circle at 5, shade left*   (B) *Closed circle at 5, shade left*

(C) *Open circle at 5, shade right*   (D) *Closed circle at 5, shade right*

18. *The table below shows how a savings account grows each week. Write the equation that relates savings s to weeks w, and predict the savings after 8 weeks.*

Weeks (w)	Savings (s)
0	$20
1	$25
2	$30
3	$35
4	$40

Your Answer:

19. *A triangle and a rectangle both have a base of 6 cm and a height of 10 cm. How do their areas compare?*

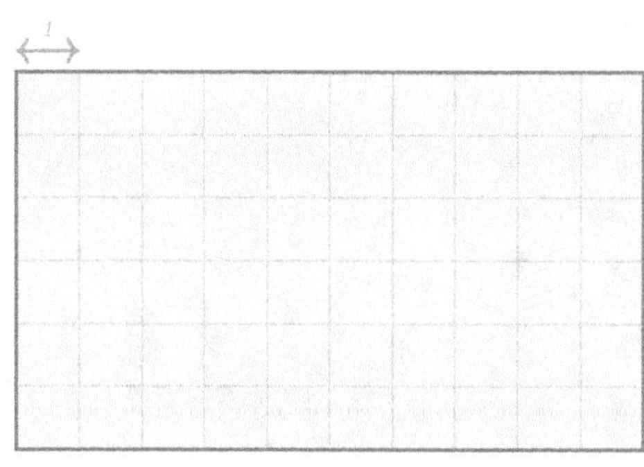

(A) The triangle has twice the area of the rectangle.

(B) They have the same area.

(C) The triangle has half the area of the rectangle.

(D) The rectangle has three times the area of the triangle.

20. *A cube has edges of 4 cm. What is the volume?*

(A) $12\ cm^3$

(B) $16\ cm^3$

(C) $48\ cm^3$

(D) $64\ cm^3$

21. *A rectangle has vertices $(0,0)$, $(6,0)$, $(6,4)$, and $(0,4)$. What is the area?*

(A) 10 square units

(B) 20 square units

(C) 24 square units

(D) 12 square units

22. *Which pair of dimensions gives a rectangular prism with the greatest surface area?*

(A) $2 \times 3 \times 10$

(B) $4 \times 4 \times 4$

(C) $1 \times 5 \times 8$

(D) $3 \times 3 \times 6$

23. A small circle has a radius of 5 cm and a large circle has a radius of 10 cm. How much greater is the area of the large circle than the area of the small circle? Use $\pi \approx 3.14$.

Your Answer:

24. Look at the flowchart below. Which path correctly identifies a statistical question?

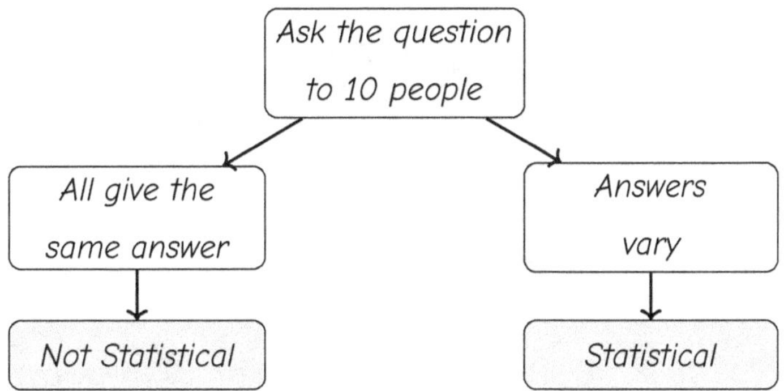

Using the flowchart, which question would end at "Statistical"?

(A) How many wheels does a bicycle have?

(B) What is the sum of $10 + 5$?

(C) How many minutes do you spend eating lunch?

(D) How many days are in one week?

25. If every value in a data set is increased by 5, what happens to the range?

(A) It increases by 5.

(B) It decreases by 5.

(C) It stays the same.

(D) It doubles.

26. A histogram shows minutes of exercise: 0–14 (3), 15–29 (8), 30–44 (12), 45–59 (7). How many people exercised less than 30 minutes?

Your Answer:

27. In a box plot, the "box" stretches from —

(A) the minimum to the maximum   (B) Q1 to Q3

(C) the median to the maximum   (D) 0 to the median

28. Two box plots are shown on the same number line.

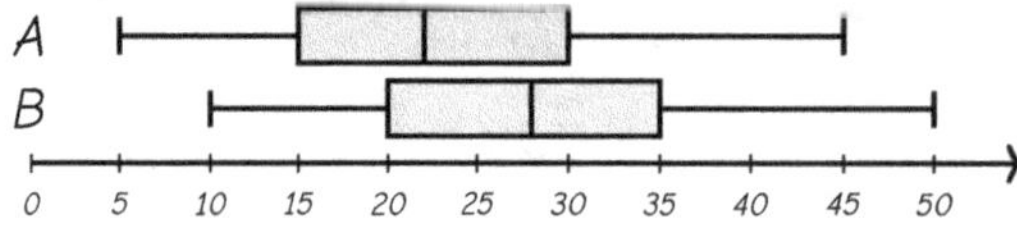

Which data set has a greater IQR?

(A) A (IQR = 15)   (B) B (IQR = 15)

(C) They have the same IQR.   (D) Cannot be determined.

29. A bag has 5 orange balls, 3 purple balls, and 2 yellow balls. What is the probability of **not** drawing a yellow ball?

(A) $\dfrac{2}{10}$ (B) $\dfrac{5}{10}$

(C) $\dfrac{8}{10}$ (D) $\dfrac{3}{10}$

30. A circle graph shows pets owned by 120 students: Dogs 40%, Cats 30%, Fish 15%, Birds 15%. How many students own cats?

(A) 30   (B) 36

(C) 40   (D) 48

 # End of Practice Test 10 

Great job finishing the test!

📋 My Score

I got _____________ out of 30 questions right.

Check your answers in the **Answer Key** at the back of the book.

💡 Review any questions you missed. That's how we learn!

📊 Check Your Score Online!

Visit **ViewMath Academy** to enter your answers and see which topics you need to review. You can also explore lessons, take quizzes, track your scores, and save your progress!

viewmath.com/score/6.1.MN.25

Or go to viewmath.com/score and enter code: 6.1.MN.25

Answer Key & Explanations

Answer Key

First try each test on your own, then check your work here.

✓ Practice Test 1 — Answer Key

 5    C    B    B    C

 Ella has a proportional relationship. After 6 lawns, Ella earns \$90 and Jake earns \$82, so Ella earns more.

 A    A    15   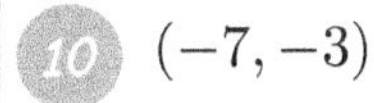 $(-7, -3)$  B    $A = 4, B = (x + 3)$  22    B

 Answers vary. Example: You buy n notebooks at \$5 each and pay \$3 for shipping.    B

 No. Both shade to the left, but $x < 8$ has an open circle at 8 and $x \leq 8$ has a closed circle at 8.

 $y = 3, y = 11, y = 19$  C    B    B    A    C    B    D

 B    B   28 Not always.   29 B   30 \$450

💡 Time to Learn! 💡

*Review the explanations below, **especially for the questions you missed**.*

Understanding why each answer is correct builds stronger problem-solving skills.

Tip: *Circle any questions you got wrong, then read their explanation carefully.*

📖 Practice Test 1 — Detailed Explanations

1. Each part $= 20 \div 4 = 5$. Vegetables $= 1 \times 5 = 5$.

2. From the graph, each hour earns \$10 (2 units on the y-axis $=$ \$10). In 6 hours: $6 \times 10 = \$60$.

3. Ratio: $1 : 4$. When $x = 3$: $y = 3 \times 4 = 12$.

4. $0.15 \times 100 = 15\%$.

5. The table shows that only the credit card does not use your own money and can charge interest, meaning it involves borrowing.

6. Ella: $y = 15x$ is proportional ($k = 15$). Jake: $y = 12x + 10$ is non-proportional ($b = 10 \neq 0$). For $x = 6$: Ella earns $15 \times 6 = 90$; Jake earns $12 \times 6 + 10 = 82$. Ella earns more.

7. The number line shows 5 jumps of $\dfrac{1}{6}$ to reach $\dfrac{5}{6}$. This models $\dfrac{5}{6} \div \dfrac{1}{6} = 5$.

8. Multiply: 5.4×6. As whole numbers: $54 \times 6 = 324$. One decimal place: 32.4 liters.

9. $|-9| = 9$ and $|-6| = 6$. Adding these gives $9 + 6 = 15$. Find each absolute value first, then add.

10. Reflecting across the x-axis changes the sign of the y-coordinate: $(x, y) \to (x, -y)$. So $(-7, 3)$ becomes $(-7, -3)$.

11. Sandwiches cost $6s$ and drinks cost $2d$. Total: $6s + 2d$.

12. $4(x + 3) = 4 \times (x + 3)$. The two factors are 4 and $(x + 3)$.

Find more at
ViewMath.com/MN-Grade6

ViewMath.com

13 $3(8+2) - 8 = 3(10) - 8 = 30 - 8 = 22.$

14 *Distribute:* $3 \times x + 3 \times 5 = 3x + 15.$

15 $5n$ *represents a cost per item and 3 represents a fixed fee.*

16 *"Below 0" means less than 0:* $t < 0.$ *(Not equal to 0, since it says "below.")*

17 *The direction is the same, but the circle type differs: open vs. closed.*

18 $4(1) - 1 = 3.\ 4(3) - 1 = 11.\ 4(5) - 1 = 19.$

19 *One sail:* $\frac{1}{2} \times 3 \times 7 = 10.5\ m^2.$ *Two sails:* $10.5 \times 2 = 21\ m^2.$

20 *Box A volume* $= 6 \times 4 \times 5 = 120.$ *Box B:* $120 = 10 \times 3 \times h = 30h,$ *so* $h = 4.$

21 *Base* $= |2 - (-4)| = 6.$ *Height* $= |5 - 0| = 5.$ *Area* $= \frac{1}{2} \times 6 \times 5 = 15$ *square units.*

22 $SA = 2(12)(8) + 2(12)(5) + 2(8)(5) = 192 + 120 + 80 = 392\ in^2.$

23 $A = \pi r^2 = 3.14 \times 5^2 = 3.14 \times 25 = 78.5\ in^2.$ *Choice B uses the circumference formula* $2\pi r.$ *Choice D uses* $2\pi r^2.$

24 *The answers range from 0 to 4 and vary from person to person, which confirms it is a statistical question.*

25 *Range* $= 35 - 5 = 30.$

26 1 hour has 6 dots (highest frequency) ⊠ mode. 8 hours is far from the cluster (0–4) and has only 1 dot ⊠ outlier.

27 The left whisker extends from the minimum to Q1, covering the lowest 25% of the data.

28 A single outlier can make the range very large while the rest of the data is tightly clustered. The IQR gives a better picture of overall spread. For example, $\{1, 50, 51, 52, 53\}$ has range 52 but most values are close together.

29 Numbers ≤ 3: $1, 2, 3$. That is 3 out of 6. $P(\leq 3) = \dfrac{3}{6} = \dfrac{1}{2}$.

30 $15\% \times 3{,}000 = 0.15 \times 3{,}000 = 450$. The family spends \$450 on entertainment.

☑ Practice Test 2 — Answer Key

1 C

2 12, 16, and 30

3 Part A: Runner A $= 3:4$; Runner B $= 2:3$.   Part B: Runner A is faster.

4 C

5 C

6 C

7 B

8 \$16.43

9 B

10 The point $(9, 0)$ lies on the x-axis because its y-coordinate is 0. Points on the axes are not inside any quadrant.

11 C

12 5 and $(y + 8)$

13 30 square cm

14 $6a + 10$

15 A

16 C

17 B

18 C

19 $54\ in^2$

20 A

21 A

22 $268\ m^2$

23 B

24 No

25 C

26 Total $= 100$, Mode $=$ Bus, Walk $= 15\%$

27 B

28 A

29 C

30 C

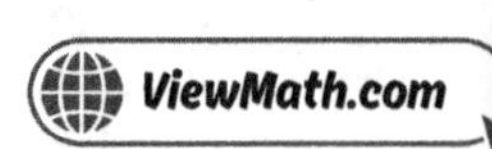

💡 *Time to Learn!* 💡

*Review the explanations below, **especially for the questions you missed**.*

Understanding why each answer is correct builds stronger problem-solving skills.

***Tip:** Circle any questions you got wrong, then read their explanation carefully.*

📖 Practice Test 2 — Detailed Explanations

1. *Girls have 6 parts = 24, so each part = $24 \div 6 = 4$. Boys have 4 parts = $4 \times 4 = 16$.*

2. *Row 2: $4 \times 2 = 8$, so $6 \times 2 = 12$. Row 3: $6 \times 4 = 24$, so $4 \times 4 = 16$. Row 4: $4 \times 5 = 20$, so $6 \times 5 = 30$.*

3. *Runner A: 3 laps in 4 min = 0.75 laps/min. Runner B: 2 laps in 3 min $\approx$ 0.667 laps/min. $0.75 > 0.667$, so Runner A is faster. On the graph, Runner A's line is steeper.*

4. $0.20 \times \$25 = \$5.$

5. *Savings: $0.20 \times \$240 = \$48$. Supplies: $0.10 \times \$240 = \$24$. Left: $\$240 - \$48 - \$24 = \$168$.*

6. *The ratio flour:sugar $= 2 : 3$. Multiply both by 3: $6 : 9$. You need 9 cups of sugar. This is a proportional relationship.*

7. $\dfrac{3}{5} \times \dfrac{5}{1} = \dfrac{15}{5} = 3.$

8. $12.95 + 3.48 = 16.43.$ *Hundredths: $5 + 8 = 13$, write 3, carry 1. Tenths: $9 + 4 + 1 = 14$, write 4, carry 1. Ones: $2 + 3 + 1 = 6$. Tens: 1.*

Find more at
ViewMath.com/MN-Grade6

9. Absolute value tells you how far a number is from zero. -12 is 12 units from zero, so $|-12| = 12$. A common mistake is to think $|-12| = -12$, but distance is never negative.

10. The four quadrants are the regions between the axes. A point is on an axis when $x = 0$ (on the y-axis) or $y = 0$ (on the x-axis). Since $(9, 0)$ has $y = 0$, it sits on the x-axis itself, which is the boundary between Quadrant I and Quadrant IV—not in either one.

11. Equally distributing c cupcakes into 6 boxes means dividing: $c \div 6$.

12. $5(y + 8) = 5 \times (y + 8)$. The factors are 5 and $(y + 8)$.

13. $A = \frac{1}{2} \times 10 \times 6 = \frac{1}{2} \times 60 = 30$ square cm.

14. $3a + 5a - 2a = 6a$ and $7 + 3 = 10$. Result: $6a + 10$.

15. Tickets: $9p$. One bucket of popcorn: 6. Total: $9p + 6$.

16. "Fewer than 20" means less than 20: $y < 20$.

17. Both shade to the right. The only difference is the circle at 5: open for $>$, closed for $\geq$.

18. Each dog has 4 legs: $L = 4d$.

19. For a right triangle the legs are the base and height. $A = \frac{1}{2} \times 9 \times 12 = 54$ in^2.

20. Base area $= 10 \times 4 = 40$. Height $= 120 \div 40 = 3$ cm.

21. Length $= |6 - (-3)| = 9$. Width $= 54 \div 9 = 6$ units.

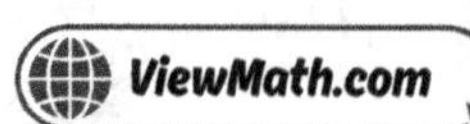

22. $SA = 2(10)(8) + 2(10)(3) + 2(8)(3) = 160 + 60 + 48 = 268\ m^2.$

23. First find the radius: $r = 12 \div 2 = 6\ m.$ Then $A = \pi r^2 = 3.14 \times 6^2 = 3.14 \times 36 = 113.04\ m^2.$ Choice D uses the diameter instead of the radius in πd^2.

24. "How many inches in a foot?" has a numerical answer (12) but is not statistical because the answer does not vary. A statistical question must produce data that varies.

25. $IQR = Q3 - Q1 = 55 - 35 = 20.$

26. $45 + 30 + 15 + 10 = 100.$ Bus has the highest frequency ⮕ mode. Walk: $15/100 = 15\%.$

27. The five-number summary consists of minimum, first quartile (Q1), median, third quartile (Q3), and maximum.

28. Same medians, but Athlete A's range (1.2) is smaller — the times vary less, indicating more consistency.

29. Event Y is positioned closest to 0.75 on the probability scale.

30. All sectors must total 100%. $100 - 40 - 25 - 15 = 20\%.$

📋 Practice Test 3 — Answer Key

1 C	**2** C	**3** B	**4** C	**5** C	**6** B	**7** A	**8** B	**9** C	**10** B
11 D	**12** 1	**13** C	**14** $8y + 24$	**15** B	**16** C	**17** D	**18** C	**19** B	
20 B	**21** A	**22** $108\ cm^2$	**23** B	**24** B	**25** B	**26** $Mode = 5000, Range = 4000$			
27 B	**28** B	**29** C	**30** C						

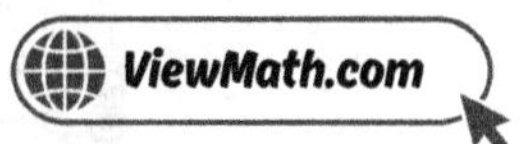

> 💡 *Time to Learn!* 💡
>
> *Review the explanations below, **especially for the questions you missed**.*
>
> *Understanding why each answer is correct builds stronger problem-solving skills.*
>
> ***Tip:*** *Circle any questions you got wrong, then read their explanation carefully.*

📖 Practice Test 3 — Detailed Explanations

1. *The question asks flour to eggs. Flour $= 3$, eggs $= 2$. The ratio is $3 : 2$.*

2. *$6 \times 3 = 18$ cups of flour, so $4 \times 3 = 12$ eggs.*

3. *The ratio is $6 : 2 = 3 : 1$. When $x = 15$: $y = 15 \div 3 = 5$.*

4. *Discount: $0.15 \times \$80 = \$12$. Sale price: $\$80 - \$12 = \$68$.*

5. *Interest $= \$500 \times 0.18 = \$90$. New balance $= \$500 + \$90 = \$590$.*

6. *A proportional relationship has the form $y = kx$ with no added or subtracted constant. Only $y = 7x$ fits this form.*

7. *$\dfrac{1}{3} \times \dfrac{3}{2} = \dfrac{3}{6} = \dfrac{1}{2}$.*

8. *Multiply as whole numbers: $12 \times 3 = 36$. Count decimal places: 1.2 has 1 and 0.3 has 1, total 2. Place the decimal: 0.36.*

9. *$|6| = 6$ and $|-6| = 6$. Both 6 and -6 are 6 units from zero, so both have an absolute value of 6.*

Find more at
ViewMath.com/MN-Grade6

ViewMath.com

10 In Quadrant II, the x-coordinate is negative (left of the y-axis) and the y-coordinate is positive (above the x-axis), giving the sign convention $(-, +)$.

11 Giving away 6 means subtracting: $x - 6$.

12 $4(a + b)$ is a single term — it is one product. (If you distribute it to $4a + 4b$, it would be 2 terms.)

13 $3^2 + 3 = 9 + 3 = 12$.

14 $8 \times y + 8 \times 3 = 8y + 24$.

15 $d = 60t$ means distance equals 60 miles per hour times the number of hours.

16 "At most 21" means 21 or fewer: $d \leq 21$.

17 Closed circle at -2 (included) with shading to the left: $x \leq -2$.

18 $c = 7(6) = 42$ dollars.

19 $15 \times 4 = 60$ is the rectangle area. Sam forgot to multiply by $\frac{1}{2}$. The correct area is 30 cm^2.

20 $V = 3 \times 2 \times 2 = 12$ ft^3.

21 Rectangle area: $7 \times 3 = 21$. Triangle area: $\frac{1}{2} \times 2 \times 3 = 3$. Remaining: $21 - 3 = 18$ square units.

22 2 triangles: $2 \times \frac{1}{2}(3)(4) = 12$ cm^2. 3 rectangles: $8 \times 3 + 8 \times 4 + 8 \times 5 = 24 + 32 + 40 = 96$ cm^2. Total: $12 + 96 = 108$ cm^2.

Find more at
ViewMath.com/MN-Grade6

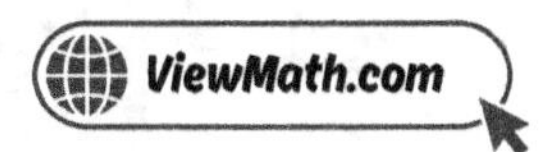

23 The curved part is half the circumference: $\frac{2\pi r}{2} = \pi r = 3.14 \times 7 = 21.98$ cm. The straight edge equals the diameter: $d = 2 \times 7 = 14$ cm. Total perimeter $= 21.98 + 14 = 35.98$ cm. Choice A forgets the straight edge.

24 There are always 3 feet in one yard. This is not a statistical question because the answer does not vary.

25 MAD (Mean Absolute Deviation) is calculated by finding the average of the distances of each data value from the mean.

26 5000 has frequency 6 (highest). Range $= 7000 - 3000 = 4000$.

27 10 values. Median $= (9 + 11) \div 2 = 10$. Lower half: $1, 3, 5, 7, 9$. $Q1 = 5$. Upper half: $11, 13, 15, 17, 19$. $Q3 = 15$. $IQR = 15 - 5 = 10$.

28 A small IQR means the middle 50% is tightly packed. A large range means the minimum and maximum are far apart, indicating extreme values stretch the data.

29 There are 6 equally likely outcomes. Only 1 of them is a 4. $P(4) = \frac{1}{6}$.

30 Savings $+$ Charity $= 30\% + 10\% = 40\%$. Then $40\% \times \$40 = 0.40 \times 40 = \$16$.

✅ Practice Test 4 — Answer Key

1 B   2 64   3 C

4 Yes. $95 at 10% off $= \$85.50$. $105 at 20% off $= \$84$. Adding the item saves $1.50.    5 C    6 D

7 A   8 1.44   9 C   10 B   11 A   12 7 and 4   13 B   14 $11m + 2$ 15 B

16 C   17 B   18 B   19 B   20 A   21 B   22 C   23 C

Find more at
ViewMath.com/MN-Grade6

24 The teacher has one specific age, so the answer does not vary.   **25** C   **26** B   **27** B   **28** A

29 A: 0 (impossible); B: $\frac{1}{2}$ (equally likely); C: $\frac{3}{4}$ (likely); D: 1 (certain)   **30** B

💡 **Time to Learn!** 💡

Review the explanations below, **especially for the questions you missed.**

Understanding why each answer is correct builds stronger problem-solving skills.

Tip: Circle any questions you got wrong, then read their explanation carefully.

📖 Practice Test 4 — Detailed Explanations

1 "3 pencils for every 1 eraser" means pencils to erasers is $3 : 1$.

2 $3 \times 8 = 24$ red, so white $= 5 \times 8 = 40$. Total $= 24 + 40 = 64$.

3 $5 : 15$ simplifies to $1 : 3$.

4 At \$95: 10% off gives $\$95 - \$9.50 = \$85.50$. At \$105: 20% off gives $\$105 - \$21 = \$84$. $\$84 < \$85.50$, so yes.

5 Groceries are food, which is a basic need. The other choices are things you might enjoy but can live without.

6 A proportional relationship has the form $y = kx$. When $x = 0$, $y = 0$, so the line always passes through the origin $(0, 0)$.

7 Dividing by $\frac{c}{d}$ is the same as multiplying by its reciprocal, $\frac{d}{c}$.

Find more at
ViewMath.com/MN-Grade6

8. Multiply as whole numbers: $36 \times 4 = 144$. Count decimal places: $1 + 1 = 2$. Place the decimal: 1.44. Check: $3.6 \times 0.4 = 1.44$.

9. Distance from sea level is measured by absolute value. $|-40| = 40$ and $|25| = 25$. Since $40 > 25$, the diver is farther from sea level. The sign tells direction (above or below), not distance.

10. Quadrant II has sign convention $(-, +)$. Since $-4 < 0$ and $7 > 0$, the point $(-4, 7)$ is in Quadrant II. A common mistake is choosing Quadrant III, which has both coordinates negative.

11. $\frac{n}{3}$ is "n divided by 3." Adding 7 gives "7 more than n divided by 3."

12. 7 and 4 are the terms without variables, so they are constants.

13. $P = 2(7) + 2(3) = 14 + 6 = 20$ cm.

14. Distribute: $8m + 2 + 3m$. Combine: $8m + 3m = 11m$. Result: $11m + 2$.

15. A square has 4 equal sides, so its perimeter is 4 times the side length: $P = 4s$.

16. "Below freezing" means below $32°F$: $t < 32$.

17. $x < -3$: $-4 < -3$ is true (shaded), but $-2 < -3$ is false (not shaded).

18. $21 = 3x$, so $x = 21 \div 3 = 7$.

19. $P = \frac{1}{2} \times 10 \times 6 = 30$. $Q = \frac{1}{2} \times 10 \times 8 = 40$. Difference $= 40 - 30 = 10$ cm^2.

20. $V = \frac{1}{2} \times \frac{1}{2} \times \frac{1}{2} = \frac{1}{8}$ m^3.

Find more at
ViewMath.com/MN-Grade6

21 $Base = |8 - 2| = 6.$ $Height = |7 - 1| = 6.$ $Area = \frac{1}{2} \times 6 \times 6 = 18$ square units.

22 2 triangular bases $+3$ rectangular faces $= 5$ faces total.

23 Area of $P = \pi \times 4^2 = 16\pi.$ Area of $Q = \pi \times 8^2 = 64\pi.$ Ratio $= 64\pi \div 16\pi = 4.$ When the radius doubles, the area quadruples because area depends on r^2.

24 A statistical question expects data that varies. The teacher's age is a single fixed number.

25 Upper half: $12, 14, 16.$ The median of the upper half is 14, so $Q3 = 14.$

26 $Range = 74 - 68 = 6.$

27 Short whiskers and a narrow box indicate small range and small IQR, meaning data values are close together.

28 $Q_1 = (15 + 18)/2 = 16.5.$ $Q_3 = (25 + 28)/2 = 26.5.$ $IQR = 26.5 - 16.5 = 10.$

29 Event A: A standard die has faces 1–6, so rolling a 7 is impossible; $P = 0.$ Event B: $P(tails) = \frac{1}{2}.$ Event C: $P(red) = \frac{6}{8} = \frac{3}{4} = 0.75.$ Event D: All faces (1–6) are less than 7, so $P = 1$ (certain).

30 $100\% - 28\% - 34\% = 38\%.$ Candidate C received 38% of the votes.

☑ Practice Test 5 — Answer Key

1 10   **2** B   **3** B   **4** C   **5** C

6 (a) Proportional   (b) Non-proportional   (c) Proportional   **7** C   **8** C   **9** B   **10** $(3, -2)$

Find more at
ViewMath.com/MN-Grade6

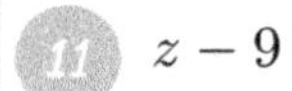 $z - 9$  D 13 25 14 B 15 B

16 $n = 4$ is NOT a solution to $n > 4$ (since 4 is not greater than 4). $n = 4$ IS a solution to $n \geq 4$ (since 4 equals 4)

17 B 18 B 19 C 20 B 21 64 square units 22 B 23 B 24 B 25 A

26 C 27 Range $= 55$, $IQR = 30$ 28 B 29 (a) $\dfrac{3}{8}$ (b) $\dfrac{1}{4}$ (c) $\dfrac{5}{8}$ 30 C

💡 Time to Learn! 💡

*Review the explanations below, **especially for the questions you missed**.*

Understanding why each answer is correct builds stronger problem-solving skills.

Tip: *Circle any questions you got wrong, then read their explanation carefully.*

📖 Practice Test 5 — Detailed Explanations

1. Total parts $= 3 + 2 = 5$. Each part $= 25 \div 5 = 5$. Swimming $= 2 \times 5 = 10$.

2. $2 : 3$ multiplied by 3 gives $6 : 9$. The other pairs do not simplify to $2 : 3$.

3. Ratio $= 3 : 2$. Double: $(6, 4)$. Triple: $(9, 6)$, not $(9, 4)$ or $(9, 8)$.

4. Savings $= \$40 - \$30 = \$10$. $\$10 \div \$40 = 0.25 = 25\%$.

5. A credit card lets you borrow money from the card company. If you do not pay on time, you are charged interest.

6. (a) $y = 9x$ is in the form $y = kx$ with $k = 9$. (b) $y = 4x + 6$ has $+6$, so it is non-proportional. (c) $y = \frac{1}{2}x$ is in the form $y = kx$ with $k = \frac{1}{2}$.

Find more at
ViewMath.com/MN-Grade6

7 $\frac{2}{3} \times \frac{5}{4} = \frac{10}{12} = \frac{5}{6}$.

8   0.6 has 1 decimal place and 0.7 has 1 decimal place. Add them: $1 + 1 = 2$ decimal places. The product is 0.42.

9 $|-10| = 10$ because -10 is 10 units from zero. Absolute value is never negative. A common mistake is to think $|-10| = -10$, but distance from zero is always positive or zero.

10   Start at $(0,0)$. Move 4 left and 3 up: $(-4,3)$. Then move 7 right: $-4 + 7 = 3$, and 5 down: $3 - 5 = -2$. The final point is $(3, -2)$.

11   "9 fewer than z" means subtract 9 from z: $z - 9$.

12 $6(n+2)$ means $6 \times (n+2)$. The two factors being multiplied are 6 and $(n+2)$.

13 $4^2 + 2(4) + 1 = 16 + 8 + 1 = 25$.

14   Distribute: $4 \times 3y - 4 \times 2 = 12y - 8$.

15   Notebooks cost $3n$ and pens cost $1 \times p = p$. Total: $3n + p$.

16 $>$ does not include equality. $\geq$ includes equality.

17 $x > 2$ does NOT include 2, so the circle should be open, not closed.

18   Height increases by 3 cm per week: $h = 3w$.

19 $A = \frac{1}{2} \times 5 \times 12 = 30$ in^2.

Find more at
ViewMath.com/MN-Grade6

20 $V = l \times w \times h$. If h becomes $2h$, then $V_{new} = l \times w \times 2h = 2(lwh)$, so the volume doubles.

21 Bottom rectangle: $10 \times 4 = 40$. Top-left rectangle: $6 \times (8 - 4) = 6 \times 4 = 24$. Total: $40 + 24 = 64$ square units.

22 $SA = 2(10)(3) + 2(10)(7) + 2(3)(7) = 60 + 140 + 42 = 242 \ m^2$.

23 $C = \pi d$, so $d = C \div \pi = 31.4 \div 3.14 = 10$ cm. Choice A gives the radius instead of the diameter.

24 Different sixth graders spend different amounts of time on homework, so the answers vary. The other questions all have a single fixed answer.

25 The mean is 10. Every value equals 10, so every distance from the mean is 0. $MAD = 0 \div 5 = 0$.

26 Since two values share the highest frequency of 4, the data set is bimodal — it has 2 modes.

27 $Range = 80 - 25 = 55$. $IQR = 65 - 35 = 30$.

28 Same median (same typical sales). Store A's range (\$200) is much smaller than Store B's (\$800), making Store A more consistent.

29 There are 8 equal sections: 3 red, 2 blue, 3 green. (a) $P(red) = \dfrac{3}{8}$. (b) $P(blue) = \dfrac{2}{8} = \dfrac{1}{4}$. (c) $P(not\ green) = 1 - \dfrac{3}{8} = \dfrac{5}{8}$.

30 $40\% - 25\% = 15\%$. Then $15\% \times 600 = 0.15 \times 600 = 90$ more students in Grade 7.

📋 Practice Test 6 — Answer Key

Find more at
ViewMath.com/MN-Grade6

1. Bananas to yogurt: $3:4$; Yogurt to bananas: $4:3$ 2. B   3. A   4. C   5. B   6. C

7. C   8. C   9. D   10. C   11. C   12. C   13. 16   14. B   15. C   16. C

17. $x \leq -4$ 18. $b = s \div 50$ (or $b = \dfrac{s}{50}$)   19. B   20. $60\ cm^3$ 21. B   22. $348\ in^2$ 23. C

24. C   25. B   26. C   27. D   28. B   29. C   30. C

💡 Time to Learn! 💡

Review the explanations below, **especially for the questions you missed**.

Understanding why each answer is correct builds stronger problem-solving skills.

Tip: Circle any questions you got wrong, then read their explanation carefully.

📖 Practice Test 6 — Detailed Explanations

1. "3 bananas for every 4 cups of yogurt" gives $3:4$. Flip the order for yogurt to bananas: $4:3$.

2. $3 \times 2 = 6$, so $5 \times 2 = 10$.

3. The ratio is $2:3$. Choice A continues with $6:9 = 2:3$. Choice B has $6:8$ which is not $2:3$.

4. Discount: $0.25 \times \$60 = \$15$. Sale price: $\$60 - \$15 = \$45$.

5. Saving regularly builds a fund for emergencies and future goals like college, a car, or unexpected bills.

6. In $y = kx$, the value k is called the constant of proportionality. It tells how many units y changes for each unit of x.

7. $4 \div \dfrac{2}{3} = \dfrac{4}{1} \times \dfrac{3}{2} = \dfrac{12}{2} = 6$ batches.

8. Move the decimal one place in both numbers: $93.6 \div 12 = 7.8$. Check: $7.8 \times 1.2 = 9.36$.

9. The opposite of -3 is 3. Both -3 and 3 are 3 units from zero, but on opposite sides of the number line.

10. The x-coordinate -3 tells you to move 3 units to the left, and the y-coordinate 4 tells you to move 4 units up. Choice D swaps the coordinates.

11. Half of n is $\frac{n}{2}$. "6 more" means add 6: $\frac{n}{2} + 6$.

12. Part C points to 3, which is a term with no variable — a constant.

13. $8^2 = 64$. Then $64 \div 4 = 16$.

14. Distribute: $3x + 12 + 2x$. Combine: $3x + 2x = 5x$. Result: $5x + 12$.

15. Hourly earnings: $8h$. Plus tip: $8h + 15$.

16. "More than \$25" means greater than 25: $d > 25$.

17. Closed circle means $\leq$ or $\geq$. Shading left means less than or equal to: $x \leq -4$.

18. Divide total students by 50 per bus: $b = s \div 50$.

19. $A = \frac{1}{2} \times 12 \times 9 = 54$ ft^2.

20. $V = 4 \times 3 \times 5 = 60$ cm^3. Each 1-cm cube has volume 1 cm^3.

Find more at
ViewMath.com/MN-Grade6

ViewMath.com

21. Width $= 56 \div 8 = 7$ units.

22. $SA = 2(15)(6) + 2(15)(4) + 2(6)(4) = 180 + 120 + 48 = 348$ in^2.

23. $A = \pi r^2 = 3.14 \times 9^2 = 3.14 \times 81 = 254.34$ m^2. Choice A uses πr and choice B uses the circumference $2\pi r$. Choice D uses $2\pi r^2$.

24. Exercise time varies from student to student. The other questions each have one fixed answer.

25. Distances from 24: $4, 2, 0, 2, 4$. $MAD = (4 + 2 + 0 + 2 + 4) \div 5 = 12 \div 5 = 2.4$.

26. Favorite color is categorical, not numerical. A frequency table or bar graph handles categories; histograms and dot plots need numerical data.

27. Different data sets can have the same five-number summary. The five-number summary only captures five key values, not every data point.

28. Week 2 median (9,500) is higher. Week 2 IQR (1,200) is smaller, meaning more consistent daily step counts.

29. A fair coin has 2 equally likely outcomes (heads or tails). $P(\text{heads}) = \dfrac{1}{2}$.

30. $40\% \times 200 = 0.40 \times 200 = 80$. So 80 students take the bus.

✅ Practice Test 7 — Answer Key

1. 6 2. B 3. $(2, 5)$ and $(8, 20)$ (or any equivalent pair) 4. C

5. Advantage: You can buy something now even if you do not have enough cash. Disadvantage: If you do not pay the balan

Find more at
ViewMath.com/MN-Grade6

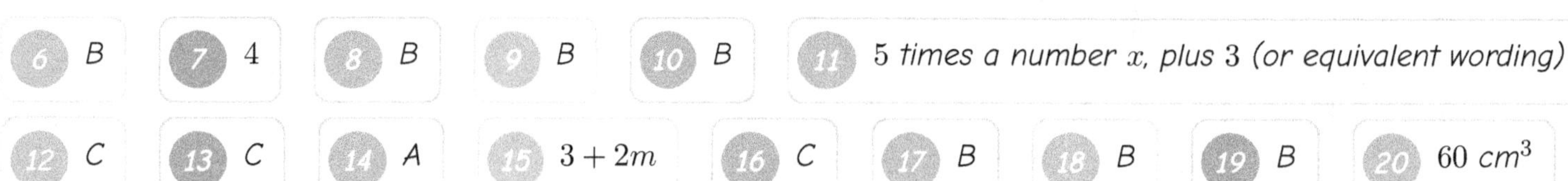

6 B **7** 4 **8** B **9** B **10** B **11** 5 times a number x, plus 3 (or equivalent wording)

12 C **13** C **14** A **15** $3 + 2m$ **16** C **17** B **18** B **19** B **20** 60 cm^3

21 B **22** B **23** 47.1 cm **24** B **25** 20 **26** C

27 Store B has higher typical sales (median $\approx$ \$275 vs. \$250). Store A has more predictable sales (IQR $\approx$ \$100 vs.

28 B **29** B **30** C

💡 Time to Learn! 💡

Review the explanations below, **especially for the questions you missed.**

Understanding why each answer is correct builds stronger problem-solving skills.

Tip: Circle any questions you got wrong, then read their explanation carefully.

📖 Practice Test 7 — Detailed Explanations

1 $21 \div 7 = 3$, so multiply both by 3: iced tea $= 2 \times 3 = 6$.

2 $2 \times 4 = 8$ and $3 \times 4 = 12$, so $8 : 12$ is equivalent to $2 : 3$.

3 Ratio $= 4 : 10 = 2 : 5$. Half gives $(2, 5)$; double gives $(8, 20)$.

4 100% means "the whole thing." 100% of $n = 1 \times n = n$.

5 Credit cards allow purchases without immediate cash but can lead to debt through interest charges if balances are not paid in full.

Find more at
ViewMath.com/MN-Grade6

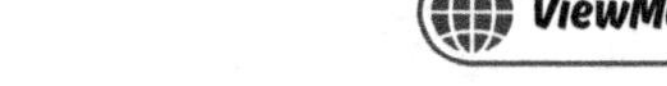

6   In choice B, $\frac{5}{1} = \frac{10}{2} = \frac{15}{3} = 5$. The ratio is constant, so it is proportional. The other tables have different ratios.

7 $2\frac{1}{2} = \frac{5}{2}$. Then $\frac{5}{2} \div \frac{5}{8} = \frac{5}{2} \times \frac{8}{5} = \frac{40}{10} = 4$ shelves.

8   Multiply: 2.75×3. As whole numbers: $275 \times 3 = 825$. Two decimal places: $8.25.

9   The opposite of a number is the same distance from zero on the other side of the number line. The opposite of 5 is -5 because both are 5 units from zero.

10   When the y-coordinate is 0, the point lies on the x-axis. The point $(-6, 0)$ is 6 units to the left of the origin on the x-axis. A common mistake is thinking a negative x-value means the point is on the y-axis.

11 $5x$ means "5 times x" and $+3$ means "plus 3."

12 $5m - 2n + 7 + m$ has 4 terms: $5m$, $2n$, 7, and m.

13 $2(9) + 2(4) = 18 + 8 = 26$.

14   Like terms: $5p + 2p + 4p = 11p$. Constants: $3 - 1 = 2$. Result: $11p + 2$.

15   Flat fee: $3. Per-mile cost: $2m$. Total: $3 + 2m$.

16   "At most 50" means 50 or less: $w \leq 50$.

17 $>$ does not include 4, so use an open circle to show 4 is NOT a solution.

18 $y = 4(3) = 12$. The point $(3, 12)$ satisfies $y = 4x$.

Find more at
ViewMath.com/MN-Grade6

19 The height is the perpendicular distance from the base to the opposite vertex. It is not always a side of the triangle.

20 $V = 2.5 \times 4 \times 6 = 60\ cm^3$.

21 Bottom rectangle: $8 \times 3 = 24$. Left rectangle above: $4 \times (7 - 3) = 4 \times 4 = 16$. Total: 40. This matches option B.

22 $SA = 2(4)(4) + 2(4)(9) + 2(4)(9) = 32 + 72 + 72 = 176\ cm^2$.

23 $C = \pi d = 3.14 \times 15 = 47.1\ cm$.

24 The dot plot shows data ranging from 0 to 5, with different students giving different answers. This variability confirms the question is statistical.

25 Median $= 115$. Q1: median of $100, 105, 110 = 105$. Q3: median of $120, 125, 130 = 125$. IQR $= 125 - 105 = 20$.

26 Every value is 5, so 5 is the value that appears most often. The mode is 5.

27 Store A: median $\approx \$250$, IQR $\approx 300 - 200 = \$100$. Store B: median $\approx \$275$, IQR $\approx 350 - 225 = \$125$. Store B sells more on a typical day, but Store A's sales are more consistent (smaller IQR and range).

28 The median describes a typical value (center) and the IQR describes how spread out the middle 50% of values are (spread). Together they summarize the data well.

29 $\frac{2}{5} = 0.4$ and $\frac{3}{4} = 0.75$. Since $0.75 > 0.4$, Event B is more likely.

30 $75\% = \frac{75}{100} = \frac{3}{4}$ after dividing the numerator and denominator by 25.

Find more at
ViewMath.com/MN-Grade6

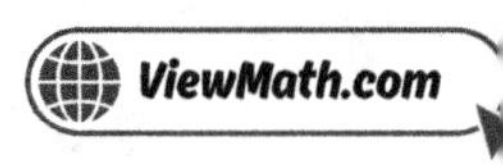

✅ Practice Test 8 — Answer Key

1 B   **2** Part A: 2 : 1;   Part B: 4 cups of sugar

3 False. Equivalent ratios always form a straight line through the origin.   **4** B

5 Interest: \$96. Total: \$896.   **6** C   **7** B   **8** B   **9** C

10 Part A: Quadrant II;   Part B: $(8, 1)$, Quadrant I;   Part C: $(-8, -1)$, Quadrant III   **11** B   **12** B

13 \$23   **14** $6x + 12$ **15** $n + 12 - 5$ or $n + 7$ **16** $g \leq 4$

17 $x = 3$ IS a solution to $x \leq 3$ because $3 \leq 3$ is true. $x = 3$ is NOT a solution to $x < 3$ because $3 < 3$ is false.

18 C   **19** 13.5 m^2 **20** B   **21** 24 square units   **22** C   **23** B

24 Answers vary. Example: "How many glasses of water do you drink each day?"   **25** 6   **26** B

27 50%   **28** C   **29** 20%   **30** B

💡 Time to Learn! 💡

Review the explanations below, **especially for the questions you missed**.

Understanding why each answer is correct builds stronger problem-solving skills.

Tip: Circle any questions you got wrong, then read their explanation carefully.

📖 Practice Test 8 — Detailed Explanations

1 Dogs have 4 parts. Each part = 2 animals. Dogs = $4 \times 2 = 8$.

Find more at
ViewMath.com/MN-Grade6

2 Part A: From the graph, $(2,1)$ shows 2 cups of flour for every 1 cup of sugar. Part B: Following the pattern, $(8,4)$, so 4 cups of sugar.

3 When you multiply both parts of a ratio by the same number, the points increase at a constant rate, which produces a straight line through $(0,0)$.

4 $35\% = \dfrac{35}{100}$. GCF of 35 and 100 is 5: $\dfrac{35 \div 5}{100 \div 5} = \dfrac{7}{20}$

5 $I = 800 \times 0.06 \times 2 = \$96$. Total $= \$800 + \$96 = \$896$.

6 Relationship P passes through $(0,0)$ with constant ratio $\frac{3}{2} = 1.5$, so it is proportional. Relationship R starts at $(0,2)$, not the origin, so it is non-proportional.

7 $\dfrac{2}{7} \times \dfrac{6}{5} = \dfrac{12}{35}$. Kai is correct.

8 Multiply as whole numbers: $12 \times 5 = 60$. Count decimal places: 0.12 has 2 and 0.5 has 1, total 3. Place the decimal: $0.060 = 0.06$.

9 The opposite of 0 is 0 itself. Zero is neither positive nor negative and sits right in the middle of the number line. It is the only number that is its own opposite.

10 Part A: $(-8,1)$ has signs $(-,+)$, which is Quadrant II. Part B: Reflecting across the y-axis flips the x-sign: $(-8,1) \to (8,1)$. Signs $(+,+)$ is Quadrant I. Part C: Reflecting across the x-axis flips the y-sign: $(-8,1) \to (-8,-1)$. Signs $(-,-)$ is Quadrant III.

11 "8 less than a number" means subtract 8 from n: $n - 8$.

12 The term d means $1 \cdot d$, so the coefficient is 1. (The subtraction sign belongs to the operation, not the coefficient of d itself in the original expression.)

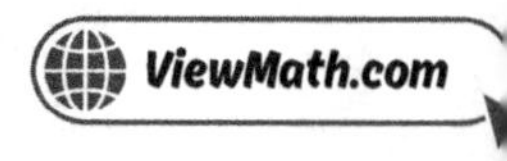

13 $5 + 3(6) = 5 + 18 = 23$ dollars.

14 Perimeter $= 2(2x + 5) + 2(x + 1) = 4x + 10 + 2x + 2 = 6x + 12$.

15 Start with n, add 12, subtract 5: $n + 12 - 5 = n + 7$.

16 "No more than 4" means 4 or fewer. $g \leq 4$.

17 $\leq$ includes the boundary value; $<$ does not.

18 You choose x (independent), and y is computed from it (dependent). y increases as x increases.

19 $A = \frac{1}{2} \times 4.5 \times 6 = 13.5\ m^2$.

20 Box A: $V = 5 \times 2 \times 3 = 30$. Box B: $V = 3 \times 3 \times 4 = 36$. Box B is greater.

21 Base $= |5 - (-3)| = 8$. Height $= |4 - (-2)| = 6$. Area $= \frac{1}{2} \times 8 \times 6 = 24$ square units.

22 A cube has 6 equal faces. $SA = 6 \times 5^2 = 6 \times 25 = 150\ in^2$.

23 $C = 2\pi r = 2 \times 3.14 \times 7 = 43.96$ cm. Choice A forgets the factor of 2. Choice D uses the area formula πr^2 by mistake.

24 Any question about food where different classmates would give different answers is acceptable.

25 Distances from 60: $10, 5, 0, 5, 10$. MAD $= (10 + 5 + 0 + 5 + 10) \div 5 = 30 \div 5 = 6$.

Find more at
ViewMath.com/MN-Grade6

26   *Favorite fruit is categorical data (not numerical intervals). A frequency table counts each category and is the natural first step.*

27   *Q1 to Q3 is the box, which always contains 50% of the data.*

28   *The median and IQR are resistant to outliers. The mean and range can be greatly affected by extreme values.*

29   *Total marbles:* $4 + 6 + 10 = 20.$ *$P(green) = \dfrac{4}{20} = \dfrac{1}{5} = 0.20 = 20\%.$*

30   *Let T be the total. $15\% \times T = 450$, so $T = 450 \div 0.15 = 3{,}000.$ The total donations were \$3,000.*

✅ Practice Test 9 — Answer Key

 C    C    A    A    C    C    D

8   Part A: 27.54 km;   Part B: 9.18 km per day    D

10   Part A: $A = (2, 5)$, $B = (5, 1)$, $C = (2, 1)$;   Part B: $A' = (-2, 5)$, $B' = (-5, 1)$, $C' = (-2, 1)$;   Part C: Quadrant

 B    B   13 A   14 A   15 C   16 C   17 D   18 C   19 C   20 B

21 B   22 B   23 B   24 C   25 B   26 B   27 A   28 B   29 B   30 75

💡 Time to Learn! 💡

*Review the explanations below, **especially for the questions you missed**.*

Understanding why each answer is correct builds stronger problem-solving skills.

Tip: *Circle any questions you got wrong, then read their explanation carefully.*

Find more at
ViewMath.com/MN-Grade6

🌐 **ViewMath.com**

📖 Practice Test 9 — Detailed Explanations

1. Total parts $= 1 + 3 + 2 = 6$. Each part $= 18 \div 6 = 3$. Blue $= 3 \times 3 = 9$.

2. $1 \times 3 = 3$ red, so $4 \times 3 = 12$ blue. Multiply both parts by 3.

3. Line A rises 4 for every 1 it moves right; Line B rises 3. A higher rise means steeper.

4. Tax: $0.06 \times \$50 = \$3$. Total: $\$50 + \$3 = \$53$.

5. He saves $\$80 - \$50 = \$30$. Percent saved $= \frac{30}{80} \times 100 = 37.5\%$.

6. $k = \frac{y}{x}$. Check: $\frac{8}{2} = 4$, $\frac{12}{3} = 4$, $\frac{20}{5} = 4$. So $k = 4$.

7. $\frac{4}{5} \div \frac{2}{5} = \frac{4}{5} \times \frac{5}{2} = \frac{20}{10} = 2$ servings.

8. Part A: $8.76 + 12.30 + 6.48 = 27.54$ km. Part B: $27.54 \div 3 = 9.18$ km per day. Check: $9.18 \times 3 = 27.54$.

9. A number and its opposite are the same distance from zero on different sides, so they cancel each other out. For example, $6 + (-6) = 0$ and $-9 + 9 = 0$.

10. Part A: Reading from the grid, $A = (2, 5)$, $B = (5, 1)$, $C = (2, 1)$. Part B: Reflecting across the y-axis changes the sign of each x-coordinate: $A' = (-2, 5)$, $B' = (-5, 1)$, $C' = (-2, 1)$. Part C: All reflected vertices have negative x-coordinates and positive y-coordinates, which is the sign convention $(-, +)$ for Quadrant II.

11. The product of 4 and t is $4t$. The sum of 9 and that product is $9 + 4t$.

12. The terms are $4x$, 9, and $2y$. Terms are separated by $+$ or $-$ signs.

Find more at
ViewMath.com/MN-Grade6

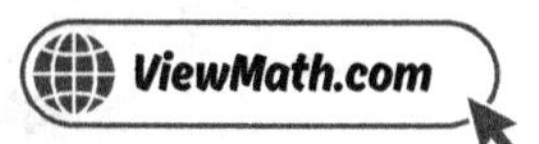

13 $6(4) - 4^2 = 24 - 16 = 8.$

14 $7n$ and $4n$ are like terms. Add coefficients: $7 + 4 = 11$, so $7n + 4n = 11n.$

15 s represents the side length of any square. It can be any positive number.

16 $n \le 8$ means n is 8 or less. $8 \le 8$ is true. 8.5, 9, and 10 are all greater than 8.

17 $10 < 10$ is false. 10 is NOT less than 10, so it is not a solution.

18 $d = 60(3) = 180$ miles.

19 $A = \frac{1}{2} \times 20 \times 7 = 70$ $in^2.$

20 Base area $= 6 \times 4 = 24$ in^2. Height $= 72 \div 24 = 3$ in.

21 Base $= |5 - 1| = 4$. Height $= |9 - 1| = 8$. Area $= \frac{1}{2} \times 4 \times 8 = 16$ square units.

22 $SA = 2(6)(4) + 2(6)(3) + 2(4)(3) = 48 + 36 + 24 = 108$ $cm^2.$

23 A semicircle is half a circle. $A = \dfrac{\pi r^2}{2} = \dfrac{3.14 \times 10^2}{2} = \dfrac{314}{2} = 157$ cm^2. Choice C is the full circle area.

24 $12 \times 9 = 108$ is a single fixed answer. It does not vary from person to person.

25 Data Set A: range $= 9 - 5 = 4$. Data Set B: range $= 10 - 4 = 6$. Data Set B has the larger range.

26 The size of the data set and whether it is numerical or categorical determine the best display. Categorical ⊠ frequency table/bar graph. Small numerical ⊠ dot plot. Large numerical ⊠ histogram.

Find more at
ViewMath.com/MN-Grade6

27 Class A's box (IQR) is narrower than Class B's. A narrower box means the middle 50% of scores are closer together — more consistent.

28 The range uses only the maximum and minimum. Changing any extreme value changes the range. However, the IQR only uses Q_1 and Q_3 and can stay the same even if extremes change. Still, the range is most directly influenced by outliers since it depends on the two most extreme values.

29 $P(cherry) = \dfrac{4}{10}$ and $P(grape) = \dfrac{4}{10}$. The probabilities are equal, so these events are equally likely.

30 Swimming $= 100\% - 40\% - 35\% = 25\%$. Then $25\% \times 300 = 0.25 \times 300 = 75$ students.

✅ Practice Test 10 — Answer Key

1 C   **2** B   **3** C   **4** B   **5** B

6 Part A: Ana; $y = 2x$.   Part B: Ben's line starts at $(0, 1)$, not the origin, so his relationship is non-proportional.

7 $\dfrac{7}{4}$ or $1\dfrac{3}{4}$ **8** 21.6   **9** D   **10** D   **11** $4x + 5$ **12** B   **13** $A = \$14,\ B = \$23$

14 A   **15** B   **16** $a \geq 16$ **17** B   **18** $s = 5w + 20$; after 8 weeks: $60   **19** C   **20** D

21 C   **22** A   **23** 235.5 cm²   **24** C   **25** C   **26** 11   **27** B   **28** C   **29** C

30 B

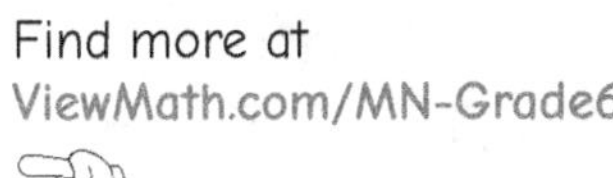
Find more at
ViewMath.com/MN-Grade6

💡 Time to Learn! 💡

*Review the explanations below, **especially for the questions you missed**.*

Understanding why each answer is correct builds stronger problem-solving skills.

Tip: *Circle any questions you got wrong, then read their explanation carefully.*

📖 Practice Test 10 — Detailed Explanations

1 Apple has 3 parts, orange has 5 parts. Total parts = 8. Total ounces = $8 \times 4 = 32$.

2 The ratio is $2 : 50$. For 6 bags: $50 \times 3 = 150$ sq ft.

3 At \$2 per bottle, 4 bottles cost \$8, not \$10. The point $(4, 10)$ does not fit the ratio.

4 Sale price: $\$45 - 0.20 \times \$45 = \$45 - \$9 = \$36$. Tax: $0.08 \times \$36 = \$2.88$. Total: $\$36 + \$2.88 = \$38.88$.

5 Credit interest owed: $\$150 \times 0.10 = \$15$. Savings interest earned: $\$150 \times 0.02 = \$3$. Difference: $\$15 - \$3 = \$12$.

6 Part A: Ana's line passes through $(0,0)$ with $k = \frac{2}{1} = 2$, so $y = 2x$. Part B: Ben's line starts at $(0,1)$. Since the graph does not pass through the origin, Ben's relationship is non-proportional (equation: $y = 1.5x + 1$).

7 $\dfrac{7}{10} \times \dfrac{5}{2} = \dfrac{35}{20} = \dfrac{7}{4} = 1\dfrac{3}{4}$.

8 Move the decimal one place in both numbers: $86.4 \div 4 = 21.6$. Check: $21.6 \times 0.4 = 8.64$

9 The absolute value of 0 is 0 because zero is 0 units from itself. Zero is the only number whose absolute value is 0.

10 Quadrant IV has sign convention $(+, -)$. Since $2 > 0$ and $-5 < 0$, the point $(2, -5)$ is in Quadrant IV. A common error is confusing Quadrant IV with Quadrant II which has signs $(-, +)$.

11 There are 4 triangles (each worth x) and 5 unit squares. The expression is $4x + 5$.

12 $\frac{n}{5} = \frac{1}{5} \times n$. The factors are $\frac{1}{5}$ and n.

13 $h = 2$: $8 + 3(2) = 14$. $h = 5$: $8 + 3(5) = 23$.

14 Distribute: $7k - 21 + 5$. Combine: $-21 + 5 = -16$. Result: $7k - 16$.

15 In 4 days you read $4r$ pages. Remaining: $100 - 4r$.

16 "No fewer than 16" means 16 or more: $a \geq 16$.

17 $\leq$ means 5 is included (closed circle). Less than 5 is to the left (shade left).

18 Savings increase by \$5 per week, starting at \$20: $s = 5w + 20$. When $w = 8$: $5(8) + 20 = 60$.

19 Rectangle area $= 6 \times 10 = 60$ cm^2. Triangle area $= \frac{1}{2} \times 6 \times 10 = 30$ cm^2. The triangle is half.

20 A cube has $l = w = h = 4$. $V = 4 \times 4 \times 4 = 64$ cm^3.

21 Length $= 6$, width $= 4$. Area $= 6 \times 4 = 24$ square units.

22 A: $2(6) + 2(20) + 2(30) = 112$. B: $6(16) = 96$. C: $2(5) + 2(8) + 2(40) = 106$. D: $2(9) + 2(18) + 2(18) = 90$. Option A is greatest.

Find more at
ViewMath.com/MN-Grade6

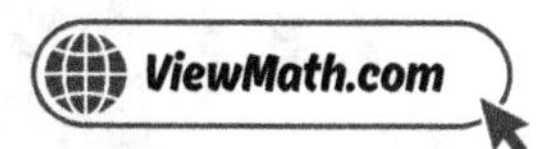

23. *Small circle:* $A = 3.14 \times 5^2 = 3.14 \times 25 = 78.5 \ cm^2$. *Large circle:* $A = 3.14 \times 10^2 = 3.14 \times 100 = 314 \ cm^2$. *Difference* $= 314 - 78.5 = 235.5 \ cm^2$.

24. *If you ask 10 people how long they spend eating lunch, the answers vary. Following the flowchart, varying answers lead to "Statistical."*

25. *Adding 5 to every value shifts the max and min by the same amount, so the difference (range) stays unchanged.*

26. *0–14: 3 people. 15–29: 8 people. Total* < 30 *min:* $3 + 8 = 11$.

27. *The box stretches from Q1 to Q3, containing the middle 50% of the data.*

28. *A: IQR* $= 30 - 15 = 15$. *B: IQR* $= 35 - 20 = 15$. *Both have IQR* $= 15$.

29. *Total balls:* $5 + 3 + 2 = 10$. $P(yellow) = \dfrac{2}{10}$. $P(not \ yellow) = 1 - \dfrac{2}{10} = \dfrac{8}{10}$.

30. $30\% \times 120 = 0.30 \times 120 = 36$. *So 36 students own cats.*

Well done checking your answers!

Keep practicing to strengthen your skills.

Find more at
ViewMath.com/MN-Grade6

ViewMath.com

Author's Final Note

I hope you enjoyed this book as much as I enjoyed writing it. Whether you are a student working through the material, a parent supporting your child's learning, or a teacher guiding your class, I have tried to make this book as clear and engaging as possible. I hope I have succeeded. If you have any suggestions for improvement, please let me know. I would love to hear from you.

The accuracy of calculations is very important to me. We have done our best, but I also expect that I have made some minor errors. Constant improvement is the name of the game. If you find any errors, please let me know. I will fix them in the next edition.

For students: Your learning journey does not end here. I have written a series of books to help you learn math. Make sure you browse through them. I especially recommend workbooks and practice tests to help you prepare for your exams.

For parents: Thank you for investing in your child's education. I encourage you to explore the companion resources available online to help support your child outside the classroom.

For teachers: Thank you for the invaluable work you do every day. I hope this book serves as a useful resource in your classroom. Feel free to reach out if you have suggestions or would like to discuss how best to use this book with your students.

I also enjoy reading your reviews. If you have a moment, please leave a review on where you found this book. It will help others find this book. If you have any questions or comments, please feel free to contact me at rNazari@ViewMath.com.

And one last thing: Remember to use online resources for additional help. I recommend using the resources on https://ViewMath.com You can find video lessons, practice problems, and more. You can also use the online companion for this book to track your progress and access additional resources.

Wishing all students the best in their studies, parents every success in supporting their children, and teachers continued inspiration in their classrooms!

Dr. A. Nazari

 # Great Job! Keep Learning with ViewMath!

*Keep up the great work! Visit **viewmath.com/MN-Grade6** for free lessons, quizzes, and more.*

Study Guide

Workbook

Step-by-Step

3 Practice Tests

5 Practice Tests

7 Practice Tests

Find more at
ViewMath.com/MN-Grade6

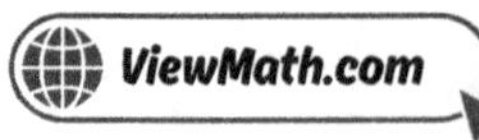